Whole Life Costing for Sustainable Building

Whole life costing is now integral to building procurement, both for new buildings and major refurbishments. It is key when assessing investment scenarios for estates as well as individual buildings, and has become a tool for justifying higher capital cost items.

Standard whole life costing methods combine capital cost, facilities costs, operational costs, income and disposal costs with a "single action–single benefit" approach. Costing based on this type of single attribute assessment misses out on realising value from the intricacies of the interactions buildings have with their occupants, users and the location in which they are placed. In contrast, the multi-attribute approach presented by the author of this book explains how to analyse the whole cost of a building, while also taking into account secondary and tertiary values of a variety of actions that are deemed important for the project owners and decision-making stakeholders. The process is an effective tool for presenting a good business case within the opportunities and constrains of real life. For example, it presents the interdependencies of how:

- Building location affects servicing strategies which impact on maintainability and control and, by extension, on occupant comfort;
- Material selection affects time on site, building maintainability as well as overall building quality and the environment;
- Building shape impacts on servicing strategies as well as operating costs.

The reader will be shown how to incorporate this method of whole life valuation into standard cost models allowing for a more robust decision making process. This is done by breaking down project aims into their most basic aspects and adopting the methods of simple quantitative risk analysis, the functionality of which is based on real data.

Written by an author immersed in project team collaboration to identify the interdependencies of design decisions throughout her professional life, this is the most practical guide available on the topic.

Mariana Trusson is a chartered engineer with over 20 years' experience in the building services and sustainability sector. She has been involved in over 400 projects at different stages of their life, from master planning and strategy development, to post-occupancy and operation. She mentors younger engineers and designers, volunteers her time and expertise for organisations such as the Chartered Institution of Building Services Engineers and regularly gives talks and lectures on designing sustainable buildings across the UK.

Whole Life Costing for Sustainable Building

Mariana Trusson

First published 2020
by Routledge
2 Park Square, Milton Park, Abingdon, Oxon OX14 4RN

and by Routledge
52 Vanderbilt Avenue, New York, NY 10017

Routledge is an imprint of the Taylor & Francis Group, an informa business

© 2020 Mariana Trusson

The right of Mariana Trusson to be identified as author of this work has been asserted by her in accordance with sections 77 and 78 of the Copyright, Designs and Patents Act 1988.

All rights reserved. No part of this book may be reprinted or reproduced or utilised in any form or by any electronic, mechanical, or other means, now known or hereafter invented, including photocopying and recording, or in any information storage or retrieval system, without permission in writing from the publishers.

Trademark notice: Product or corporate names may be trademarks or registered trademarks, and are used only for identification and explanation without intent to infringe.

British Library Cataloguing-in-Publication Data
A catalogue record for this book is available from the British Library

Library of Congress Cataloging-in-Publication Data
Names: Trusson, Mariana, author.
Title: Whole life costing for sustainable building / Mariana Trusson.
Description: Abingdon, Oxon; New York, NY: Routledge, 2020. |
Includes bibliographical references and index. |
Identifiers: LCCN 2019028978 (print) | LCCN 2019028979 (ebook) |
ISBN 9781138592582 (hbk) | ISBN 9781138775558 (pbk) |
ISBN 9781315644752 (ebk) | ISBN 9781317289968 (adobe pdf) |
ISBN 9781317289944 (mobi) | ISBN 9781317289951 (epub)
Subjects: LCSH: Building—Cost effectiveness. | Sustainable buildings—Maintenance and repair—Cost effectiveness. |
Life cycle costing.
Classification: LCC TH438.15 .T78 2020 (print) |
LCC TH438.15 (ebook) | DDC 690.068/1—dc23
LC record available at https://lccn.loc.gov/2019028978
LC ebook record available at https://lccn.loc.gov/2019028979

ISBN: 978-1-138-59258-2 (hbk)
ISBN: 978-1-138-77555-8 (pbk)
ISBN: 978-1-315-64475-2 (ebk)

Typeset in Times New Roman
by codeMantra

For Philip

Contents

Acknowledgements

Everyone I have met has in one way or another influenced the contents in this book; from the despondent facilities managers I met in passing while carrying out existing building surveys to the curious attendees of community engagement sessions for new buildings, all of them gave me an insight. However, there are some people that have been instrumental through the years in keeping me engaged and they need a special mention.

Colin Ashford gave me the seed of the idea for this book; he endured extended phone conversations on principles and provided over a thousand documents, books and presentations to serve as the start of my background reading. This would not be possible without him and his infectious passion for good building design.

I would also like to thank Ewan McGeer and Sophie Simpson for their enthusiasm and continual encouragement and Alister Philip for his level-headedness; I would not have had the audacity to complete this without them.

Finally, I want to thank the publishers for being so patient with my numerous missed deadlines, you guys are awesome.

Introduction

The most important aspect of building procurement and operation is cost. The capital cost of the building, and associated services, represents only a small part of this. The additional operating and maintenance costs of the building in many respects outweigh the initial investment several times over the building's lifetime.

With HM Treasury guidance stipulating that construction procurement choices should be made on the basis of whole life costs (WLC) and commercial organisations assessing investment opportunities based on projected return, WLC is being used on the majority of buildings. Coupled to this is the requirement for ever-more-efficient buildings and the associated increase in capital cost for the technology and design to achieve this.

As a result, there are a number of standard methods and documents on the subject of WLC including BS ISO 15686, the H M Treasury Green Book, the Office of Government Commerce (OGC) Procurement Guide 07 and BS EN 60300. A common theme in all of these is the treatment of single items (such as the building design, the building fabric, lighting and mechanical services) as having a singular cost implication in terms of capital costs and a singular implication in terms of operation and/or maintenance requirements.

This book has been written to illustrate the theory of assessing multiple associations for each cost item to include the effect systems have on each other within a functioning building, thus illustrating the true impact of design choices as they would be in real life for living, breathing buildings taken as whole interactive systems rather than singular asset items. The audience will include, I hope, building design professionals and private developers as well as public policy makers. These groups include those interested in creating policies to promote the construction and refurbishment of financially viable, comfortable, resource- and energy-efficient buildings.

The integration of a multiple cost impact assessment to standard WLC models is also necessary for developing the business case for investment outside the UK. With UK firms increasingly working and designing buildings for international clients, a methodology that can be easily adapted to any WLC model can be invaluable.

So is this yet another WLC model to confuse the already saturated field? The short answer is no. This book is not intended to provide an alternative method with new nomenclature and different parameters. It is intended to become a part of whatever existing methodology is used with the aim of enhancing the accuracy of the assessment to ensure the client makes a decision based on reasoned judgement. It is anticipated that the methodology will be used as *part* of the overall process to creating the client's business case along with the technical and environmental assessments incorporating aspects from both of these to provide the single final solution to the project.

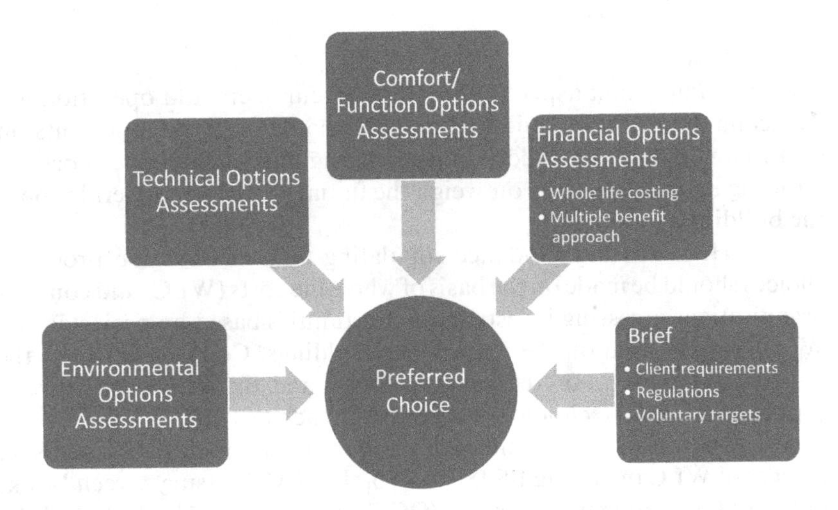

In addition, this methodology attempts to bridge the gap that exists in many client institutions between capital cost budgets and operating and maintenance cost budgets, by providing the links between the two in a transparent and detailed manner.

In this book I describe the methods needed to take into account the effect on running costs of the physical interactions between all building aspects, from orientation and positioning to lighting, façade design and occupancy patterns.

The linchpin of this approach is adaptability. This book is not intended to provide a guide to all the interactions between systems, but rather to give a methodology for incorporating additional, flexible and dynamic parameters to a rather steady state model of calculating future costs. These parameters will inevitably be different for each unique building design; however, the method for taking them into account can be used consistently.

The book is divided into three main areas. The first introduces WLC and its benefits along with a review of the different approaches used in the field. The second introduces the various "sustainable" options as incorporated in current designs and provides details on how each of these affects other (previously unrelated in cost terms) aspects of the building. The third section provides the practical aspects of costing these options using a multiple cost impact approach giving an example of how this can be achieved in practice.

The style of this book is such to make is accessible to a wide readership and so has avoided unnecessary technical jargon where possible. Where technical vocabulary has been used, it has been explained in the text. In addition, I have tried to remove any bias towards a specific geographic location to allow for the use of this method internationally; however, where ever location specific details have been necessary I have made this clear to allow for users to adjust any parameters as needed.

1 Whole life valuation

Why it is important

Looking at the big picture

In a nutshell, buildings exist to allow, among others:

- Businesses to operate in a productive environment;
- Retailers to facilitate sales;
- Governments (local and central) to promote civil necessities;
- Educational establishments to provide an effective learning environment;
- Healthcare facilities to provide a healing environment;
- Residences to provide shelter and safe havens for occupants.

There are a number of real-world business objectives that are met through effective building design. These are not traditionally associated with decisions taken during building and building services procurement; however, there are strong correlations to indicate that they should be taken into account.

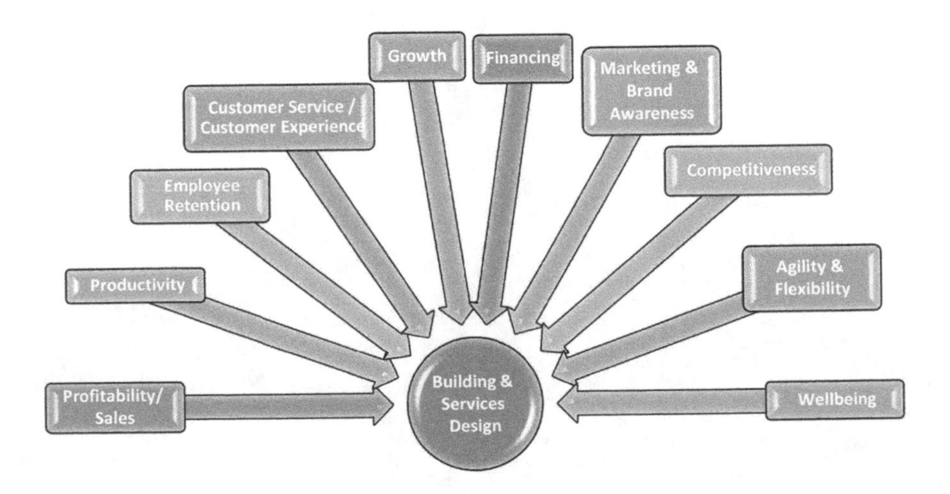

Profitability/sales prosperity

Profitability is undeniably the most important business objective. Profitability is the essence of a sustainable business model and, as such, is considered sacrosanct; and rightly so. The focus should be controlling costs in both production and operations while maintaining the profit margin on products or services traded.

In the retail sector, everything is measured by sales, which in turn are affected by a number of aspects such as brand loyalty, footfall, etc.

Productivity

Pivotal to profitability is productivity. Good business sense means providing all of the resources that employees need to remain as productive as possible. This includes training, equipment and, of course, buildings.

For governmental or not for profit organisations productivity is principal to achieving public spending key performance indicators (KPIs). Top of the list here are educational establishments where pupil/student/researcher productivity is very publicly visible through test/exam/research results.

Employee retention

It costs an average of £30k[1] to replace an employee in lost productivity and the costs associated with recruiting, which include employment advertising and paying placement agencies. Twelve percent[2] of exit interviews state stress or illness (other than sickness) as the reason for leaving. Maintaining a productive and positive employee environment improves retention.

Customer service/customer experience

Good customer service helps retain clients, generating repeat revenue. Studies[3] show that reduced staff turnover and happier employees, particularly those that are client-facing, lead to statistically significant higher levels of customer service.

For retailers, customer loyalty is paramount, and an enhanced customer experience is crucial to attracting and retaining customers.

Growth

Growth is directly affected by how company resources such as cash, equipment and personnel are managed.

Financing

Even an organisation with good cash flow needs financing contacts in the event that capital is needed for long-term projects.

Marketing and brand awareness

Marketing is about advertising; getting customer input; understanding consumer buying trends; being able to anticipate future needs and developing business partnerships that help an organisation to improve its market share.

Efficiency, sustainability, carbon savings and energy are all strong market trends that are here to stay.

Competitiveness

Any business decision that improves competitiveness in an international market, within sound financial constraints, is an incontrovertible advantage.

Agility and flexibility

Preparing for growth and creating processes that effectively deal with an ever-evolving and changing market is essential for any business.

Wellbeing

Feelings of wellbeing are fundamental to the overall health of an individual and they directly affect a multitude of other aspects, such as a person's life span. The rate of recovery from illness is directly linked to wellbeing, which in turn may ultimately reduce the healthcare burden releasing government funds for other services or reduce insurance premiums for individuals and companies.

Prioritising the intangibles

A business case puts a proposed investment decision into a strategic context. It should provide the information necessary to make an informed decision about whether to proceed with the investment and in what form. It should also be the basis against which continued funding will be compared and evaluated.

The importance of the business case in the decision-making process continues throughout the entire life cycle of an investment: from the initial decision to proceed to the decisions made to continue, modify or terminate the investment. The business case should be used to review and revalidate the investment at appropriate stages and whenever there is a significant change to the context, project or business function. As such, it is common that businesses cases are revisited and considered anew if the context changes materially during the course of the project.

In the building industry the context of a project is rarely put in terms of real benefit terms; rather, buildings are considered as single assets with a fixed monetary value which is not affected by their occupants.

This book is about informing the brief of a project and helping to include this "occupant" context within business cases made for building investment/maintenance. By using familiar tools, we will look at how to account for the benefits of multi-attribute whole life value (MWLV) that are undeniably significant but often not considered. These include:

- Reduced cost in the medium and long term;
- Reduced risk;
- Planned and budgeted maintenance control;
- Informed and standardised decision making;
- Continued compliance with regulations on procurement from the national and international legislative bodies and with best practice;
- Constructing a building that improves the occupier's business performance by designing from an asset performance requirement;
- Improved accuracy of asset service life projections;
- Aligning component selection with planned maintenance and renewal and supply chain engagement;
- Following through on whole life costing findings for overall project costs.

For suppliers, manufacturers and service providers MWLV allows them to:

- Demonstrate value for money in terms of the client's own criteria;
- Retain building value (beyond the effect of market forces);
- Support economic, social and natural resource sustainability goals.

MWLV provides a useful backbone against which to explore and assess the whole life sustainability of any programme and the subsequent outputs.

For example, it can be used to provide insight into full building utilisation; which translates to increased working capital for any organisation as well as some residential applications:

- Allowing office buildings to be used (and provide an income) outside of "standard" office hours;
- Allowing educational buildings to double as community building outside of term timetables;
- Allowing halls of residence to be used as hotels outwith term time, etc.;
- Or simply allowing buildings to be used to their full intended potential.

Using space utilisation survey results when considering new buildings on a university campus could lead to the procurement of smaller buildings in situations where existing building occupancy is consistently below 50% and overall utilisation is below 20%. In very broad terms, taking building occupiers and their behaviour into account and prioritising aspects such as productivity and flexibility could save a university as much as 50% of its building expansion masterplan budget.[4]

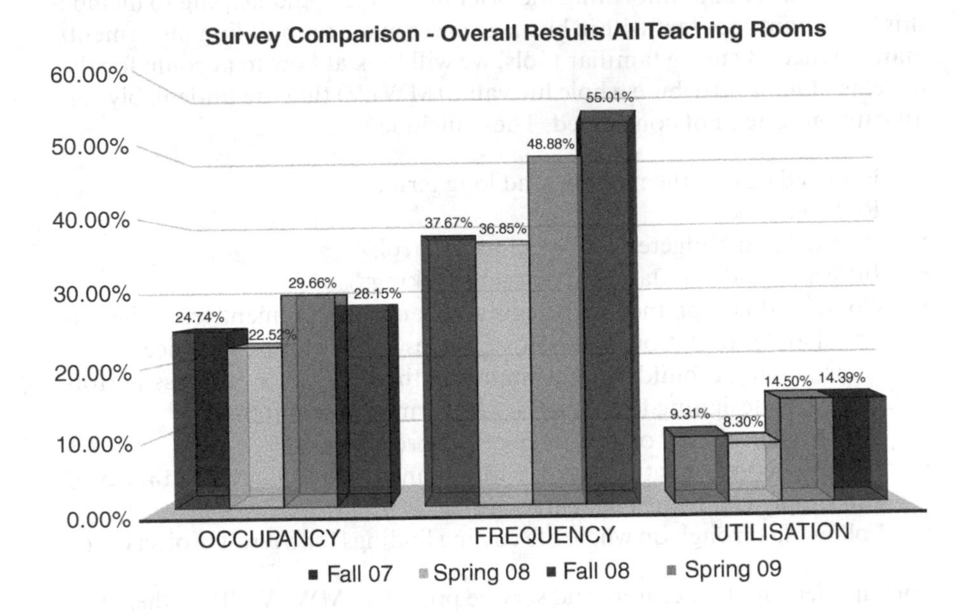

Providing accountability

The information required to develop a useful MWLV model is readily available and assumptions needn't be the basis of such an important analysis. The action of pulling reliable data together to develop the model has benefits in and of itself. For example:

- It encourages the analysis of real-life business needs;
- It provides a vessel of communication of these business needs to the project team;
- It ensures the optimisation of the total cost of ownership (TCO) and occupation by balancing capital and running costs;
- It promotes realistic short- and medium-term budgeting for operation, maintenance and repair;
- It encourages the discussions about the durability of materials and components to happen at the outset of a project;
- It helps directly confront the life expectancy of a building which currently is grossly falsely evaluated;[5]
- It provides data on actual performance and operation, compared with predicted performance, for use in future planning;
- It ensures risk and cost analysis if loss of functional performance occurs taking into account failure or inadequate maintenance;
- It helps set realistic and useful benchmarks.

What I mean by MWLV

Whole life costing

Whole life costing (WLC) is termed as the: "*economic assessment considering all agreed projected significant and relevant cost flows over a period of analysis expressed in monetary value. The projected costs are those needed to achieve defined levels of performance, including reliability, safety and availability*".[6]

In simpler terms, WLC is, according to the National Platform for Build Environment, "the systematic consideration of all relevant costs and revenues associated with the ownership of an asset". This is the definition which will be used in this book.

Total cost of ownership

The total cost of ownership (TCO) refers to the total WLC including disposal and end of life costs.

Life cycle costing

Life cycle costing (LCC) is the: "*Economic performance expressed in cost terms over the life cycle, taking account of negative costs related to energy exports and from re-use and recycling of parts of the building during its life cycle and at the end of life*".[7]

Life cycle economic balance includes LCC and adds any incomes over the life cycle and at the end of life.

Life cycle analysis

A life cycle assessment (LCA) is effectively a set of procedures. It is used for analysing the inputs and outputs of materials and energy and their associated environmental impacts directly attributable to the functioning of a product or service throughout its life-cycle. LCA is usually expressed in non-monetised terms and is focused on carbon equivalent impact.

Whole life value

Whole life value (WLV) takes into account WLC as well as functionality, social impact, quality, time and regulatory compliance. In simple terms, the WLV of an asset represents the optimum balance of stakeholders' aspirations, needs and requirements and whole life costs.[8]

The commonalities and the differences

There are slight differences between WLC and LCC. In particular LCC forms part of the WLC. And all the processes feed into whole life value to create a multi-attribute approach that can easily translate to a business case for decision makers.

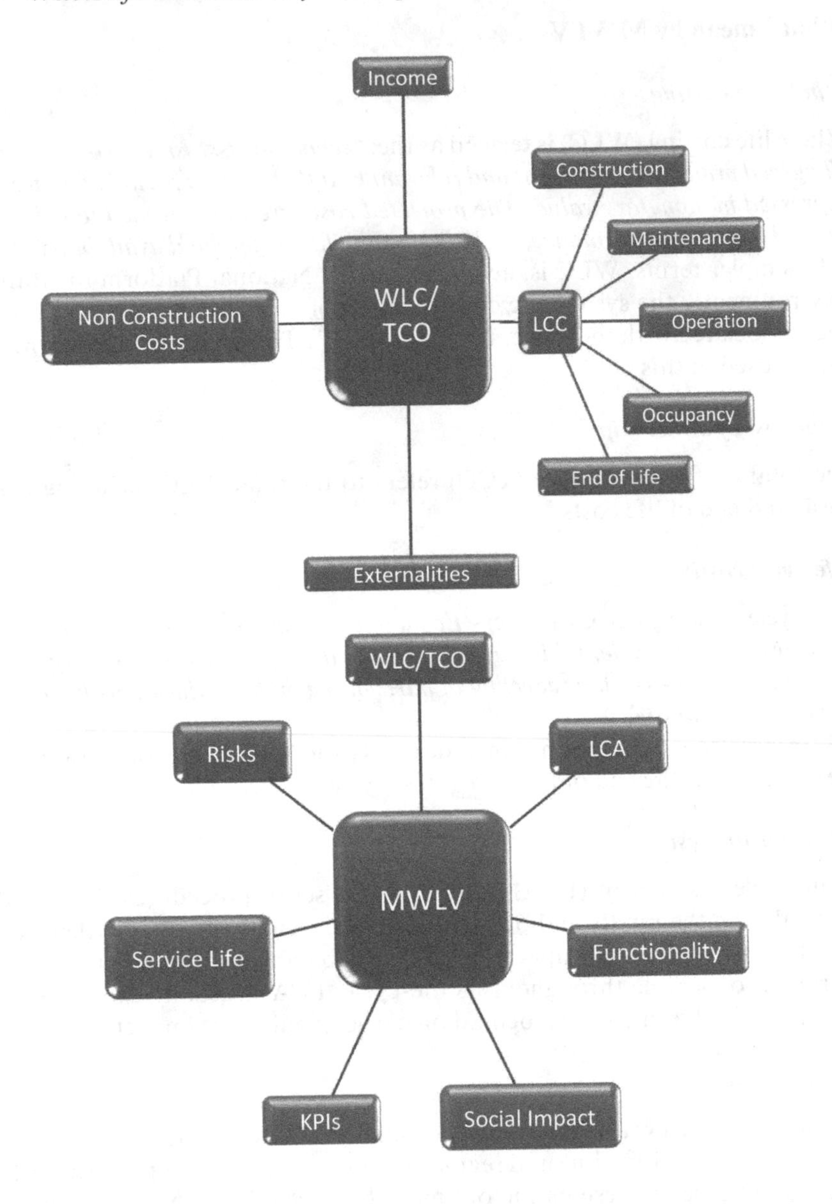

When to use MWLV

Multi-attribute whole life valuation is a tool for decision making and is most relevant when considering whole estates, whole facilities, individual buildings or structures. It is useful when comparing alternative investment scenarios such as:

- Retain and refurbish or sell or dispose;
- Alternative design;
    - Such as between concrete- or steel-framed structures.

- Alternative specifications;
  - Such as between timber and metal windows.

It is also a useful tool when items or projects have ongoing running costs; for example, lighting systems that require ongoing renewal or replacement.

MWLV modelling is vital for justifying spend within any public sector or for other initiatives where accountability and transparency is important and will be put under scrutiny. In the same vein, it is a tool for illustrating value for money during procurement.

In terms or project stages MWLV should be used at each significant stage of procurement. In particular:

- At project inception;
- At feasibility/options appraisals stages;
- At planning/design stages;
- During value engineering;
- At construction/installation/handover stages;
- During operation and maintenance;
- At end of life.

Benefits of MWLV

There are a number of key benefits associated with MWLV:

- Well-informed evaluation of competing options – MWLV is relevant to most equipment purchasing decisions, whether simple or complex and the technique is also applicable to leasing decisions;
- Better awareness of total costs – the whole life costing aspect of MWLV provides buyers and decision makers with a clear understanding of the factors governing costs and the resources required associated with each purchasing choice;
- Improved forecasting – WLC allows the full cost of a purchase over a period of time to be calculated with reasonable accuracy. It helps develop more accurate long-term cost assessment and promotes realistic budgeting for operation, maintenance and repair; this is of considerable importance when major investment decisions need to be made;
- Performance trade-offs against cost – the process provides a more holistic approach to considering value for money, the costs, quality, benefits and broader issues;
- Balances needs against preferences, thus, managing aspiration in a rational and auditable manner;
- Helps gain and maintain stakeholder support by addressing requirements at every stage;
- Creates and maintains useful and reliable benchmarks.

Weaknesses of MWLV

This analysis, as with all predictive methods, is based on assumptions. Although these are well researched at the initial stages of a project, they need to be revisited at every iteration to ensure that they are still relevant and logical.

This is a long-term analysis and we live in a rapidly developing global community. Trends; market forces; regulatory requirements; technological breakthroughs; general preferences and fashions can change many times within the serviceable life span of a project. To remain valid, realistic and reliable the WMLV model needs to be updated as significant influences change. For example, the list below indicates a few of the parameters that are likely to change within the lifespan of a building and will affect the final outcome of a model:

- Procurement/finance;
- Maintenance techniques and costs;
- Fuel costs;
- Disposal regulation and costs;
- Climate change – weather extremes;
- Resource depletion (real or just market-led);
- Safety regulations;
- Targets;
- Stakeholders.

All aspects of procurement need "*the bill payers' permission*". MWLV is a long process that in many respects needs to be undertaken by specialists. It has a cost that in many cases it can be justified by the savings identified. The analysis is complex and decisions need to be taken at many stages along the procurement, ownership and end of life journey; as such, the "*bill payers*" need to give much more than their permission; MWLV needs clear and strong buy-in. This commitment needs to extend to all applicable clients across a number of decades (the project life).

Notes

1 www.hrreview.co.uk/hr-news/recruitment/it-costs-over-30k-to-replace-a-staff-member/50677 February 2014.
2 Chartered Institute of Personnel and Development. CIPD survey report, Recruitment, retention and turnover, 2004.
3 7-2009 Unit-Level Voluntary Turnover Rates and Customer Service Quality: Implications of Group Cohesiveness, Newcomer Concentration, and Size John P. Hausknecht, Charlie O. Trevor, Michael J. Howard.
4 Durham University Space Utilisation Survey 2009.
5 Whole life value costing a new approach, Peter Caplehorn, 2012 Routledge, ISBN: 978-0-415-43423-2.
6 ISO15686-5:2008.
7 BS EN 16627:2015.
8 Achieving Whole Life Value in infrastructure and buildings, BR 476, BRE, 2005, ISBN: 1 86081 737 8.

2 Project decision making

The standard processes

Capital budgeting identifies and values potential investment opportunities to enable investors to make sound investment decisions. As such, it is undeniably an integral part of project procurement. There are a few particularities with the decisions concerning investments in the building trade, and these should be kept in mind when considering the processes laid out in this chapter.

Importantly, most options that are under consideration are mutually exclusive. For example, a building will only contain one main type of subframe at a time.

Additionally, most investment in building projects has a relatively long economically useful life and, as such, can be viewed as a long-term investment for the end users. For example, building windows have an economic life of over 20 years.

Finally, the economic life of individual building components can be vastly different from that of the building as a whole. For example, a building can be useful for over 100 years (if built appropriately) whereas the lamps installed within it may only last 5 years depending on use.

Other factors play a role as well, many of which are non-economic but which affect costs. Location is a prime example; the market identifies that some locations are "better" than others and, as such, attract higher land prices. This comparison, though, is not immediately quantifiable and, often, the factors affecting land premiums are not consistent or assessed in detail. For example, a business park in Japan may cost over 20% more to build than one in the UK. Aesthetics, branding and function are also in this category. For example, cultural or historical functions may give a building value beyond simple economic terms. And, as the theme of this book suggests, well-chosen sustainable credentials can add value; for example, over 15% of investors in Australia[1] find that there is value in certified sustainable buildings because they provide a competitive advantage making them easier to sell and lease, thus reducing losses. These same investors have already financed the development of assets worth over $40 billion in Australia.

Investment decisions

Net present value

Present value and net present value (NPV) are commonly used to form the basis for valuation for real assets and investment decisions. It is a tool used to analyse real assets with expected multi-period pay-offs. Ultimately, the method compares the cost of an investment and the present value of uncertain future cash flows generated by the project.

There are three distinct steps to calculating the NPV of a project:

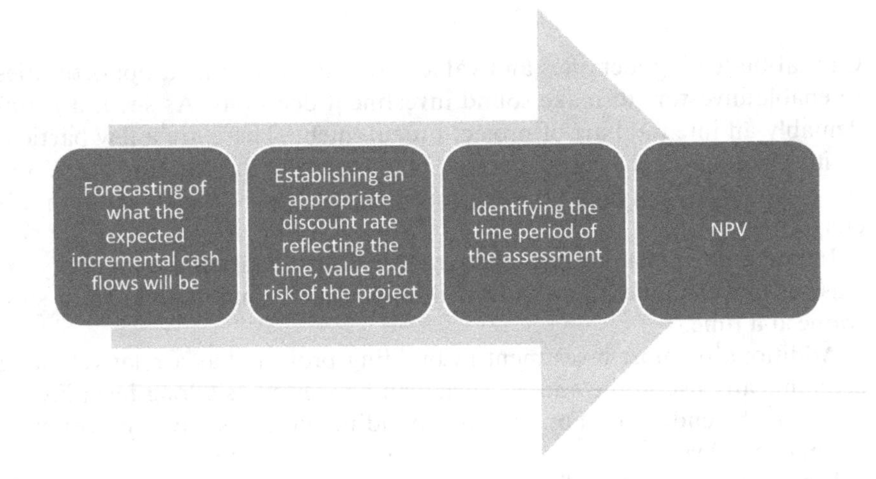

The following is the equation[2] for calculating NPV:

$$\text{NPV} = -C_0 + \frac{C_1}{(1+r)^1} + \frac{C_2}{(1+r)^2} + \frac{C_3}{(1+r)^3} \cdots \frac{C_T}{(1+r)^T} = \sum_{i=0}^{T} \frac{C_i}{(1+r)^T}$$

where:

NPV – Net present value.

C_0 – Initial cash flow (the negative sign is inserted to represent expenditure as opposed to income).

C – Forecast cash flow or the net cash income during the period (if during a time increment there is a net expenditure, this should be represented by a negative number for that period).

T – Time horizon or the number of time periods to be used in the forecast (this could be in weeks, months or years; however, it must be consistent).

r – Discount rate.

As a rule, a positive NPV should be an acceptable option as the earnings generated by the project will exceed the anticipated costs over the period. This concept is the basis for the "NPV Rule", which dictates that the only investments that should be made are those with positive NPV values.

Considerations when using NPV assessments

The tool is most appropriate for projects where all the necessary capital has been secured. Any cost of borrowing will need to be added as part of the forecast cash flow calculations and represented as a negative number.

A decision needs to be made on how to treat inflation. This can be in nominal terms or in real terms; whichever way is chosen, it needs to be consistent throughout the model(s).

When deciding on the time horizon, it is preferable to use the economic life of the asset(s). Only when this is not known should an arbitrary decision be made. If the project is not expected to continue after the end of the forecasting time horizon, then the salvage value of the asset(s) along with any tax implications needs to be included as the last expected cash flow (C_T). If, however, the project is expected to last longer than the forecasting time horizon, then a continuation value has to be established and added as the last expected cash flow (C_T).

When considering the cash flow for each time increment the following four aspects should be included:

- Operations – the cash flow generated by sales less expenses related to the operation of the project;
- Capital investments and disposal related to the project (e.g. maintenance);
- Changes to working capital – net changes in short term, assets and liabilities (e.g. energy bills);
- Tax – additional or reduced corporate tax payments associated with the project.

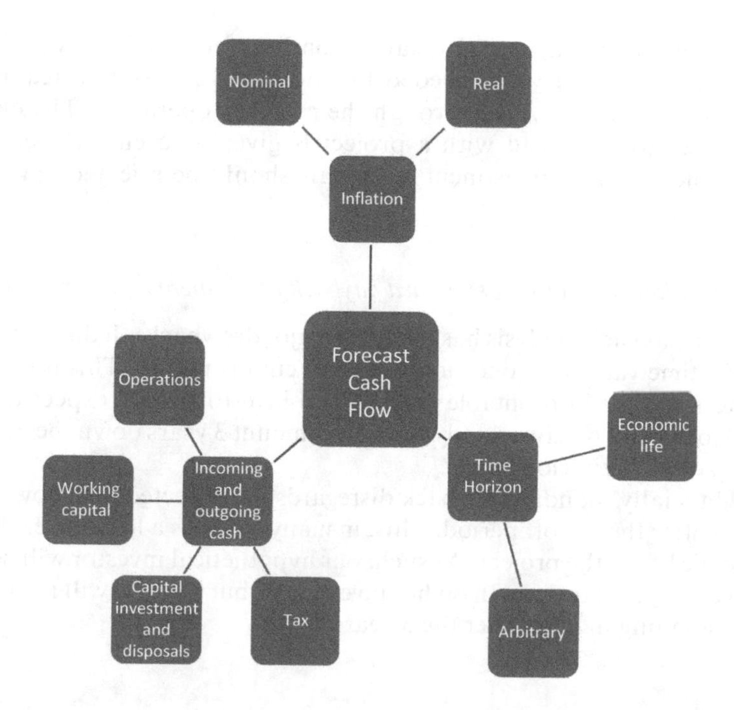

To make this tool useful the discount rate should reflect the time value[3] and risk of the project. It should include the notion of the "opportunity cost" of the capital. In simple terms, the cost of the capital should be determined by the rate of return expected from an alternative investment with a similar risk profile. This is where market expertise is required, as well as some research by the investor early on. Nonetheless, there is considerable data relating to traded financial assets that can be used to reach an informed decision with regard to the appropriate rate.

To enhance confidence in the analysis and point out major risk factors, it is good to use project risk analysis tools (e.g. assessing the risk of overheating using dynamic building simulation software).

A sensitivity analysis could also be included within the assessment. This can be used to identify key variables or value drivers for projects and helps pinpoint the most important components of the project. Thus, for example, if the NPV proves more sensitive to the working capital associated with the project than to the capital expenditure, then decision makers should focus on maximising working capital during the implementation to improve these factors.

Finally, and, perhaps, most importantly for the buildings trade, a break-even analysis should be carried out. This is a sensitivity analysis whereby each of the NPV variables is adjusted to give a calculated NPV of zero. This will greatly enhance the decision making when undertaking value engineering exercises by helping prioritise critical factors.

Standard payback

The simple payback period (years, months, weeks) of a project is the expected number of years needed for the initial investment (capital expenditure) to be recovered through the project's operation. The decision of whether to go ahead with a project is given as a cut-off period beyond which capital investment proposals should be rejected (e.g. 3, 5, 7 or 10 years).

Considerations when using standard payback assessments

Standard payback analysis has two very major drawbacks. It does not recognise the time value of money for any given cut-off period. This is extremely counterintuitive; in profitable markets, no investor would expect to invest £1000 today and receive exactly the same amount 3 years down the line, even if the payback criterion is met.

Additionally, standard payback disregards the expected cash flows from a project after the cut off period. This, in many cases, is a lot shorter than the economic life of the project. As such, our hypothetical investor will not only receive any profit or growth on her investment, but she also will not even see any continuing income after the 3 years.

Simple payback is used extensively within the buildings trade, particularly when assessing the cost effectiveness of component upgrades or replacements and component additions. Examples include window replacements, plant replacements, insulation addition, etc. In reality, simple payback is wholly unsuitable for such aspects as most building components have long term, relatively evenly distributed, cash flows that continue for much, much longer than the wholly arbitrary cut-off dates imposed by this method.

There are a few areas where simple payback analysis can form part of a sensible decision-making process; however, it is generally restricted to projects that have large payoffs in the short term and perhaps nothing later.

Discounted payback

Discounted payback is used when the time value of money is taken into account within a standard payback assessment. This is given as a present value of future cash flows within the arbitrary cut-off period.

Considerations when using discounted payback assessments

This method does not give any weight to cash flows after the cut-off period.

Discounted payback is perhaps slightly more suitable to the buildings trade, as it can take into account many of the aspects that contribute to future cash flows. Nonetheless, the cut-off period still wholly disregards the expected useful life of the components or the building. It should only really be used when comparing projects with very similar and even cash flow profiles. For example, two near identical hotels in the same location.

A good example of effective decision making with discounted payback analysis would be to compare evenly distributed rental incomes from very similar properties.

Internal rate of return

The internal rate of return (IRR) or economic rate of return (ERR) of a project is simply the discount rate (r) which returns an NPV of zero. It is effectively the rate of growth a project is likely to return.

Considerations when using IRR assessments

The rule of thumb is: the higher the IRR, the better the investment. However, there are a few circumstances where the IRR approach is not as effective as a full NPV analysis. These are:

- When the project cash flow over the period has many fluctuations between negative and positive income;
  - When positive cash flows are generated during the course of a project, the assumption is that the money will be reinvested at the project's

rate of return. However, longer term projects with fluctuating cash flows often have multiple distinct IRR values.

- When assessing mutually exclusive investment options (either due to technical or economic constraints);
- When assessing projects with different timescales.
  - For example, a project of a short duration may have a high IRR but a low overall NPV. Or a comparative may have a much longer duration and a low IRR, but an overall high NPV. The IRR assessment would favour the first project disregarding the increased value potential of the second project. Thus, while the pace at which the investment sees returns on that project may be slow, the project would be adding a great deal of overall value.

IRR is a very popular metric in estimating a project's profitability; however, it should be used in conjunction with an NPV analysis.

Profitability index (PI)

The profitability index is the ratio of the present value cash flows and the initial capital investment. Any answer returned on this ratio of 1 or over should be considered a good investment. Any value lower than 1 would indicate that the project will make a loss.

Considerations when using PI assessments

The higher the PI the more financially attractive the project is. However, PI alone is not as effective when considering mutually exclusive projects. This is because the nature of the calculation does not allow the scale of the capital cost to be compared. It also does not allow comparison or the timescale for return on the investment. Thus, again, it should be used in conjunction with a full NPV analysis.

ISO 15686-5:2008 buildings and constructed assets – service life planning – Part 5 life cycle costing

This part of ISO 15686-5 lays the basic procedures for performing LCC analyses of buildings or constructed assets and their parts. It includes for positive and negative cash flows, arising from acquisition through operation to disposal. It covers principles of LCC; instructions for LCC appraisal of options/alternatives; appraisal of LCCs in investment options; decision variables; uncertainty and risk. It also includes some worked examples. This ISO has 14 objectives and the following are most notable:

- Establish a common methodology for LCC;
- Enable the application of LCC techniques and methodology for a wide range of procurement methods;

- Address concerns over uncertainties and risks, to improve the confidence in LCC forecasting;
- Make the LCC assessments and the underlying assumptions more transparent and robust;
- Provide the framework for consistent LCC predictions and performance assessment;
- Provide a common basis for setting LCC targets during design and construction, against which actual cost performance can be tracked and assessed over the asset life span;
- Clarify the differences between LCC and WLC;
- Provide a generic menu of costs for WLC/LCC compatible with and customisable for specific international cost codes and data structure conventions.

In establishing a common methodology for LCC and clarifying the differences between LCC and WLC, ISO 15686-5 provides the detail of what should be included in LCC in a graphical form.

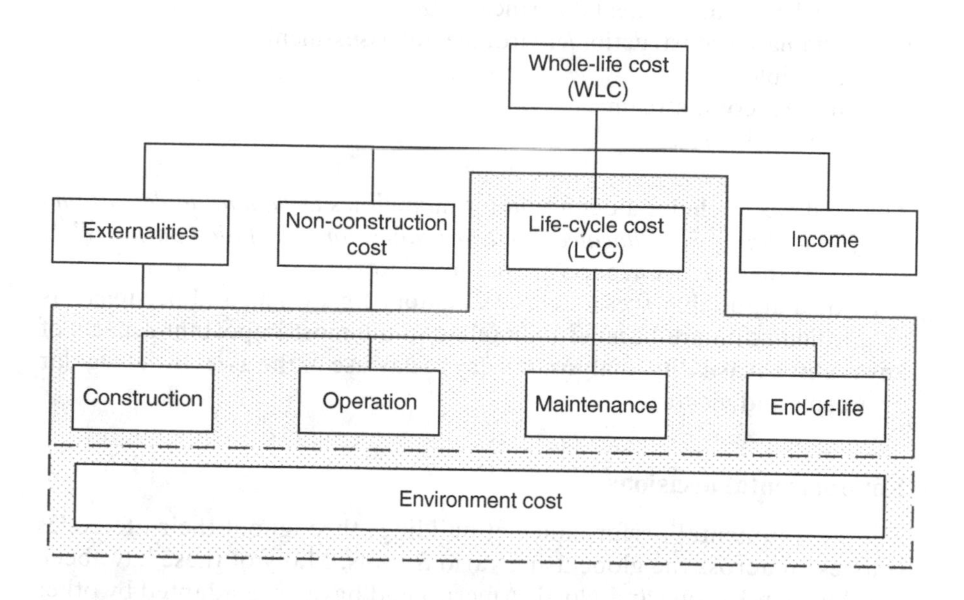

Notably, there is only a small visual connection between the design, construction and operation/maintenance of the building and the business costs.

The normal measure used in LCC analysis is indeed NPV with guidance on how to calculate real, nominal and discount costs.

Ultimately the guide provides the parameters required within the LCC as:

- Service life, life cycle and design life;
- Period of analysis;

- Cost variables for:
  - Acquisition costs;
  - Maintenance, operation and management costs;
  - Residual values/disposal costs;
  - Discount rate;
  - Inflation;
  - Taxes;
  - Utility costs including energy costs.

Considerations when following ISO 15686-5

ISO 15686-5 clearly states that the following aspects should not be included in the LCC analysis rather the WLC:

- Externalities such as the social, environmental or business costs or benefits of production and consumption;
- Environmental cost impacts;
- Social costs and benefits/sacrifices;
- Sustainable construction/environmental assessment;
- Intangibles;
- Future income streams;
- Financing costs.

The guidance includes a definition for intangibles as "*quantifiable cost and benefit that have been allocated monetary values for calculation purposes*". It states that these wider aspects are not mandatory.

I would argue that these are as mandatory for owners and occupiers as acquisition costs are because they relate to the most important aspects of any realistic feasibility analysis: the day to day use of the asset by its regular occupants and visitors.

Environmental decisions

The environmental performance of buildings throughout their life cycles is assessed across the globe using various tools. Many of these have been developed in Europe and North America and have been adapted by other regions for use. The tools are generally available for helping designers and architects to design "green" buildings. They are mainly concerned with assessing environmental impact and have the addition of LCC modelling as a small part of the process. Commonalities include the assessment of aspects such as resource efficiency, health, wellbeing and ecological protection.

Most of the assessment tools are computer-based models providing life cycle or environmental impact assessments. Some are simply methodologies to be followed during design and construction. There are estimated to be over 200 environmental assessment software packages in use. The life cycle inventory database Ecoinvent[4] is used for a number of these tools.

Assessment tools adopt one of two assessment approaches:

- Bottom-up (most common);
  - This focuses on building component, material selection, etc.
    - When using this approach tools must estimate the environmental impact of the operation phase of the building and distribute it across building materials components.
- Top-down.
  - This considers the entire building as a starting point for further improvements with initial considerations focusing on form and size and considering material and systems afterwards.

The assessment processes are either:

- Product comparison tools used during procurement;
- Decision support tools used to evaluate the building as a whole, focussing on particular aspects such as operational energy consumption or life cycle environmental impacts;
- Whole building frameworks covering a wide range of broader issues such as social and general environmental impacts.

BREEAM

The BRE Environmental Assessment Method (BREEAM[5]) is a whole building framework methodology for evaluating the environmental performance of building design. It was developed by the Buildings Research establishment in the UK and it assesses the performance of buildings in the following areas:

- Management;
- Energy use;
- Health and well-being;
- Pollution;
- Transport;
- Land use and ecology;
- Materials;
- Water.

There is also account for any Innovation activities that may be considered worthy of additional consideration within the scoring.

In general, credits are awarded by a trained assessor (and verified by the BRE) in each area according to performance against strict design target levels. A set of environmental weightings is used to produce a single overall score. It is mainly used by government agencies to ensure a level of value for money based on the credits. Developers, designers, property agents, owners and planners also use the tool as it provides a single final certificate rated on a scale of PASS, GOOD, VERY GOOD, EXCELLENT or OUTSTANDING.

Considerations when using BREEAM

BREEAM is a very popular method for "proving" sustainable credentials because of this single final certificate. It takes into account many aspects of a building from inception to occupancy and beyond. The tool has been around for over 25 years and has evolved from a niche assessment primarily targeted to commercial buildings in England to a widely recognisable "badge" of sustainability.

It has come under considerable criticism in the process. Primarily, it has been criticised for being very prescriptive. For example, the detailed actions required to award the ecology credits run to a staggering 20 pages (more once the detail of each required guidance document is taken into account). Another industry bugbear was the requirements for providing cyclist facilities which went through the minutia of detailing the size of lockers for cyclist equipment. This prescriptiveness was, and to a degree still is, considered as counterproductive because it narrows the range of design choices available to designers and consequently, inevitably, increases construction costs as suppliers capitalise on this.

In and of itself, this is not a big concern as BREEAM started out as a voluntary standard to be followed by those that were engaged enough with sustainability to seek out this tool and follow its requirements as it suits their individual projects. When used as a way to express and prove the degree of improvement over the status quo, this tool can lend additional value to a project (e.g. a BREEAM Excellent certified office building may bring in increased rent revenue than its non-rated neighbour). As such, the possible increase in construction costs to account for the BREEAM design and procurement requirements makes business sense.

Its success in this field, however, has led to its adoption as almost a mandatory process for most publicly funded projects in the UK. This move from a voluntary standard to a mandatory code across sectors such as schools and healthcare facilities has led to the greatest criticism. Laudable as it is to make sustainable buildings the norm, it misses the point of certification to prove improved performance. Design teams across the UK feel that the process tends to constrain creative thinking and channel actions into simple compliance to achieve credits rather than looking for the best outcome for each particular build (which may get no credits but may make better sense for a given situation).

This creates an inflexibility that in many cases increases costs beyond those allowed for by public bodies. For example, requiring BREEAM Very Good for each school design may be inappropriate for a school in a rural green field (poor location for achieving high BREEAM points) location without increasing budgets considerably.[6]

Percentage increases in capital costs*					
	BREEAM score (and rating) for the base case	% increase for pass	% increase for good	% increase for very good	% increase for excellent
Location					
Poor location	25.4 (pass)	0%	0%	2%	–
Typical location	39.7 (pass)	–	0%	0%	3.4%
Good location	42.2 (good)	–	–	0%	2.5%
* A small overall cost saving was identified resulting from the removal of air-conditioning equipment in the computer/server room thereby enabling increased performance at no extra cost					

The above does not account for any additional costs incurred by targeting the energy credits within BREEAM, and as such, only shows part of the picture.

It is hard for design teams and developers to make a business case for increased budgets at the best of times; but public funding is doubly scrutinised and, unless the additional "intangible" benefits of BREEAM compliance are accounted for, it is difficult to prove how BREEAM certification provides value for money.

LEED

Leadership in Energy and Environmental Design (LEED[7]) is another whole building framework methodology for evaluating the environmental performance of building design. It was developed by the United States Green Building Council (USGBC) and it assesses the performance of buildings in the following areas:

- Sustainable sites;
- Water efficiency;
- Energy and atmosphere;
- Materials and resources;
- Indoor environmental quality.

There is also account for any Innovation and Design Process activities that may be considered worthy of additional consideration within the scoring.

Credits are awarded directly by the USGBC in each area according to performance against detailed target levels. There are a number of mandatory credits that are required. It is mainly used by publicly funded agencies to ensure a level of value for money based on the credits. Investors, increasingly in the Middle East, use the tool as it also provides an internationally recognisable rate on a scale of Certified, Silver, Gold and Platinum.

Considerations when using LEED

LEED has also caused considerable discussions on whether it represents true value for money (particularly when used for publicly funded projects). There is even a website[8] devoted to LEED criticism.

Depending on a user's point of view, LEED can be more flexible than BREEAM as it is not as prescriptive on the minute details of the design and opts rather for compliance with general guidance codes such as those published by the American Society of Heating, Refrigerating and Air-Conditioning Engineers (ASHRAE[9]). Others consider LEED rigid as it does not readily allow for regional differences particularly when considering it as an assessment tool outside of the USA.

Nonetheless, LEED accreditation also has additional associated costs: The most commonly cited of these are as follows:

Leed Rating	*Potential Additional Costs*
Certified	2.5%
Silver	3.3%
Gold	5%
Platinum	8.5%

Green Star

Green Star is the Australian version of BREEAM or LEED. It has many similarities with the other two major assessment schemes both in approach and in the final certification procedure. It is managed by the Green Building Council of Australia. It assesses the performance of buildings in the following areas:

- Management;
- Indoor environment quality;
- Energy;
- Transport;
- Water;
- Materials;
- Land use and ecology;
- Emissions;
- Innovation.

The assessment is carried out by the GBCA according to performance against detailed target levels. It is mainly used by publicly funded agencies to ensure a level of value for money based on the credits. It also provides a recognisable rate on a scale of 1–6 stars.

Considerations when using Green Star

There is a considerable disconnect between the use of Green Star for existing buildings and that for new builds. In particular, less than 10% of rated Green Star buildings are refurbishments of existing buildings and none of these have achieved six stars. This is a market indication perhaps that the system may not be as viable for existing buildings and that retrofit solutions are potentially not as compliant with the criteria set.

HQE process

"La Haute Qualité Environnementale" (HQE[10]) is another whole building framework methodology for evaluating the environmental performance of building design. It was developed in France by the HQE Association, certification is operated by Cerway,[11] in all countries except France (where it is operated by the HQE association) and it assesses the performance of buildings in the following areas:

- Energy;
- Environment;
- Comfort;
- Health.

Verification of design is generally "in person" with an independent third party facilitating dialogue. It is mainly used in French-speaking countries with some uptake in other countries. Ratings are awarded by attaining the levels for each target within the assessment. Overall, HQE levels are categorised as: Pass; Good; Very Good; Excellent; Exceptional.

Considerations when using HQE

According to the France Green Building Council[12], this tool is very adaptable to a project's particular environment as it is heavily contextualised. Also, unlike BREEAM or LEED, the system is primarily focused on the overall quality of the project approach (from project management onwards) as well as in occupant comfort.

There are additional costs associated with the adoption of this tool to seek environmental credentials. Commonly it is expected that an increase in capital costs of 0.3%–10% could be incurred. Nonetheless, the system is widely used regardless of the added outlays as the additional benefits associated with occupant comfort and increase build quality have been proven in practice.

SBTool

The SBTool (formerly GBTool) is a generic building performance assessment framework for rating the sustainable performance of sites and building projects. It is spreadsheet based and has been developed by the International Initiative for a Sustainable Built Environment (iiSBE[13]). The system is mainly for use by local governments and non-government organisations (NGOs) to establish rating systems to suit their own regions and building types. It is also useful for owners and developers of large building portfolios, to specify performance requirements for developments.

The system covers a wide range of issues, but it is flexible enough to be modified in scope to include from over 115 criteria down to a minimum of 10 overall. The output of the tool is in graphical form giving the relative performance results (on a scale of 0–5) of the following issue areas:

- Site selection, project planning and development:
- Energy and resource consumption;
- Environmental loadings;
- Indoor environment quality;
- Functionality and controllability of building systems;
- Long-term performance;
- Social and economic aspects.

Considerations when using the SBTool

Rather than fixed environmental weightings, the tool includes an algorithm that automatically assigns a weighting score based on the relevance of major impact categories selected by the user. It does this by taking into account factors such as the probable intensity, duration and extent of performance effects.

The commercial rating systems such as BREEAM and LEED use a system of fixed points to give more or less importance to various issues. As such, issues may arise when the system is used outside its region of origin, and in many respects, this fixed points system is what has caused so much criticism of the tools. The SBTool takes a very different approach, as regional users can insert local context values, performance benchmarks and targets to suit certain building types.

This does require considerable client-side commitment in terms of effort and time; however, once calibrated, the system can provide much more meaningful results tailored for the individual project giving arguably better value for money.

However, if the primary objective of the exercise is not true sustainability for its own sake but includes the need for a label, then the SBTool may not be considered as appropriate.

In terms of additional capital costs, the flexibility of the tool allows assessments to be more cost neutral than other systems and a rating could be gained without additional capital expenditure needed.

CEEQUAL

The Civil Engineering Environmental Quality Assessment and Award Scheme (CEEQUAL[14]) is a credit-based assessment framework, which is applicable to civil engineering, infrastructure, landscaping and the public realm. It provides a rating for the environmental and social performance of these types of projects. It assesses the performance of buildings in thc following areas:

- Contract Strategy and Management;
- People & Communities;
- Land Use (above & below water) and Landscape;
- The Historic Environment;
- Ecology & Biodiversity;
- Water Environment (Fresh & Marine);
- Physical Resources Use & Management;
- Transport.

Much like BREEAM, CEEQUAL assessments gain a Certificate at the end of the process rated in terms of Pass, Good Very Good and Excellent.

Considerations when using CEEQUAL

The cost of carrying out a CEEQUAL assessment can be perceived to be quite high by clients. These costs are associated with application fees. There is also a large amount of time investment needed by the development team for the process with over 180 questions to be completed at the initial scoping stage.

Nonetheless, most of those that have applied CEEQUAL effectively have reported overall capital cost savings achieved both at construction and during operation.

Unlike other commercial assessment tools CEEQUAL requires full client engagement as well as considerable time contributions from the designers and contractors.

CASBEE

The Comprehensive Assessment System for Built Environment Efficiency (CASBEE)[15] is managed by the Japan Sustainable Building Consortium. The Project is assessed under the following categories for Quality and Building Environment load:

- Indoor Environment;
- Quality of Service;

- Outdoor Environment on site;
- Energy;
- Resources and Materials;
- Off-site Environment.

An award of Poor, Slightly Poor, Good, Very Good and Superior is given. The tool considers concepts such as BEE (Building Environmental Efficiency).

Each assessment item is weighted so that all the weighting coefficients within each of the assessment categories sum up to 1.0. The scores for each assessed item are multiplied by the weighting coefficient and aggregated into a total.

Considerations when using CASBEE

CASBEE is a tool that can be used only in Japan, as it has not been adapted for international use yet. It does, however, unlike most of the other listed assessment tools, include for building deconstruction.

WELL Building Standard

The WELL Building Standard[16] (WELL) is the first building standard to focus exclusively on the health and wellness of building occupants and users.

WELL takes a holistic approach to health in the built environment addressing behaviour, operations and design. Launched in October 2014 it is based on comprehensive scientific, practitioner and medical review. The standard is unique in integrating reliable research on environmental health, behavioural factors, health outcomes and demographic risk factors that affect health.

WELL also references existing standards and best practice guidelines set by governmental and professional organisations where available, in order to harmonise and clarify existing thresholds and requirements.

The standard is administered by International WELL Building Institute (IWBI) and certifies at three levels: Silver, Gold and Platinum. It is split into two major credit categories preconditions and optimisation features.

Projects must demonstrate all preconditions to achieve Silver. Projects are assessed under the following concepts:

- Air;
- Water;
- Nourishment;
- Light;
- Fitness;
- Comfort;
- Mind.

The standard is unique in linking design, operation and behavioural initiatives to the human body system.

Considerations when using WELL

The standard is very new to the market and it is expected that many aspects will be "tweaked" with increased market uptake. The approach is innovative however and it is a system that closely addresses the human aspect issues in building occupation.

Data

Reliability

Going back to basics is sometimes necessary. The reliability of the data used in valuation processes is critical and is one of the biggest factors that affect the outcome of an analysis. Reliability needs to be controlled or, at the very least, accounted for in the process.

The majority of data used for whole life value calculations is obtained by taking real life pre- and post-occupancy measurements. One of the reasons that we have not had adequate data in the past is that these measurements were not published as they were, and in some cases still are considered commercially sensitive. Another reason is that, simply, we did not know their true value. With government initiatives such as the display energy certificates in the UK and Nabers in Australia, as well as requirements by the various certification bodies mentioned in this chapter, more and more post-occupancy data is available making predictions more accurate.

Sources

Almost all reliable data sources are paid for; however, the level of accuracy allowed by these basic construction cost data is worth the price. Cost and estimation consultant firms will use their own proprietary databases for information. Nonetheless, there are sources information commercially available to all that can be used as a starting point or for verification. A number of sources are listed here, although the list is non-exhaustive and continually updated:

- Compass International (multi-national including Middle East and Russia);
- The chief estimator (Multiple currency databases);
- Dutch Association of Cost Engineers (DACE) price booklet (Netherlands, USA and Germany);
- SPON's price books (UK);
- Construction Industry Development Council (CIDC) (India);
- Rawlinsons Construction Cost Guide (Australia);

- QV Cost Builder (Quotable Value Limited) (New Zealand);
- Japan Construction Information Centre Foundation (JACIC);
- John S Page estimating manuals (USA);
- Design cost data magazine (USA);
- Richardson cost data online and Aspen capital cost estimator (USA);
- Frank R walker company (USA).

Categories

There is a considerable amount of information required for carrying out a Whole Life Value model. It can be time consuming to amass, and allowances in the programme for this collation are essential. However, the process of answering the questions, filling in the boxes or making the executive decisions in itself will bring value to the project. If anything, it will help reduce risk to the project as it considers many aspects early on.

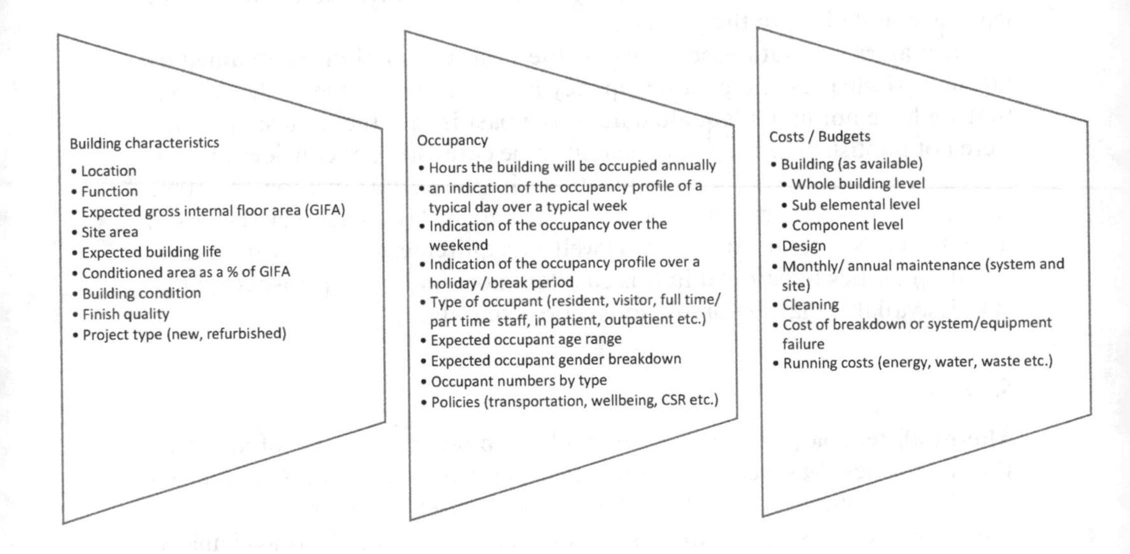

Notes

1 Valuing Green – How green buildings affect property values and getting the valuation method right, GBCA, 2008.
2 Common spreadsheet based programmes, such as Microsoft Excel and Apple Numbers, have embedded formulas for calculating NPV and IRR, the formula is included here for reference.
3 The time value of money is the concept that, provided money can earn interest, money now is worth more than the same amount tomorrow.
4 Ecoinvent is a not-for-profit association which maintains a database providing process data for thousands of products. www.ecoinvent.org/
5 www.breeam.org/

6 Sustainability costs, Building Magazine, 2005 issue 18 | By Cyril Sweet www.building.co.uk/sustainability-costs/3050602.article
7 www.usgbc.org/leed
8 www.leedexposed.com/
9 www.ashrae.org/
10 www.assohqe.org/
11 www.behqe.com/
12 www.francebgc.fr
13 www.iisbe.org/
14 www.ceequal.com
15 www.ibec.or.jp/CASBEE/english/methodE.htm
16 www.wellcertified.com/

3 Efficient design

Options under consideration

Introduction

Eighty percent of the whole life cost and environmental impacts of a building project are determined by decisions made at the design stage. A collective decision to shift our perspective from short-term gain to assessing design flaws that lead to wasted materials, money, time and effort or cause pollution has significant impacts on creating an effective building that lasts.

There are hundreds of articles available on the detrimental effects of a silo mentality. And yet buildings are still, in their overwhelming majority, designed in silos with developers, occupants, design teams, contractors, project managers, specialists and maintenance staff all working away on individual tasks born from separate goals and agendas. This mindset, when the experts involved in a project can be only focused on their own fields of expertise, is a major obstacle in delivering an effective building without compromising aesthetics, cost and profit, building longevity or the environment. As professionals in our individual fields, we end up taking routes that take us further from the ultimate goals of the project – namely, an effective building that is within budget and lasts through years of use and externalities such as climate and political change – and end up splitting the difference to "get the job done". This leads, inevitably, to mutual compromise, not fully informed agreement with regretful consequences that may not be seen for years afterwards. Sick buildings are a prime example of this.

This section aims to disperse the main barrier to achieving a mutually acceptable solution by all involved parties and by identifying aspects of building design and their interdependent effects in terms of occupant wellbeing, carbon, resource or energy savings.

Whole building

Buildings on average use 40% of total energy consumption worldwide. This is spent on heating, cooling, lighting and ventilation. The figure does not account for the embodied energy needed to develop the buildings nor does it account for the energy needed to get to the buildings (commuting, deliveries and general transportation).

Energy consumption (and subsequent carbon emissions) is affected by numerous building design decisions. The consideration of both economic and environmental aspects in building construction is paramount to achieving a balanced and efficient building design.

Building shape, envelope

The shape of a building has significant impact on the overall cost of the building as well as its energy consumption in use. Often cited as the "compactness factor", the relationship between the spatial layout of a building and its overall outline impacts greatly on aspects such as overall wall area, conditioned volume, building perimeter, building area, etc.

The following is the equation for calculating the ratio:

$$CF = \frac{A_e}{V_c}$$

where:

CF (m^{-1}) – Compactness factor, also called compactness ratio, form factor or surface to volume ratio.

A_e (m^2) – Area of building envelope.

V_c (m^3) – Volume of building conditioned space.

The lower the CF, the more compact a building is considered. It stands to reason then that the more compact a building the cheaper it will be to build and the more economic it is to run and maintain.

Interestingly, this is not a factor of open or extended discussion in building developments. Architects and cost consultants often focus on simple massing based on square meter area metrics without assessing explicitly the impacts shape has on a building cost. Typically, during these early planning and masterplanning discussions aspects such as recesses, protrusions, perimeter details which include ground beams, fascias, eaves of roofs, etc. with respect to building envelope area are not considered or addressed. But these areas soon add up particularly so in the more complex projects. As such the standard way of addressing early decision making of simply assessing the perimeter-to-floor ratio unit is not as effective in assessing building effectiveness because construction cost and overall project costs vary with plan shape complexity or irregularity.

Yet there is plenty of research worldwide indicating that shape and, in particular, compactness, can affect a building's construction cost. In essence, the higher the CF, the larger the thermal envelope of a building, which forms the greater building cost element of a construction. Increasing the CF increases build costs in almost direct proportion because it inherently means that external walls, ceilings, floors or the roof areas increase. However, when also considering and accounting for the facts that increasing the CF also increases energy consumption as well as maintenance costs, the decisions made at early stages become even more important.

The energy performance of a building is also directly proportional to its compactness factor so an increase of 1% would correlate to an equal energy consumption increase for conditioning and it's volume.

Finally, although CF and envelope shape is not the first contributor to final maintenance costs for a building, it is in the top 10 along with space allocation and orientation, with more compact building requiring lower maintenance budgets. This again is logical; less envelope area will require less upkeep compared to a building of equal footprint. A more compact building will arguably require smaller plant compared to its less compact equivalent and so on.

A building's shape, and the proportion of the building which is exposed to the environment, affects internal environmental conditions. In temperate and colder climates, the smaller the external surface area of a building the less opportunity there is for heat to escape, and in warmer climates the larger the external area the more solar gains the greater the cooling requirements for the building. Getting this balance right is only an option at the design stage of a building project.

Most of the research indicates that the most effective simple building perimeter shape is that of a square, with rectangular buildings coming close second. This is not to say that other building shapes should be abandoned, or that architects' creativity and design flair should be shunned, but that the impacts should be clearly identified at inception, development and value engineering stages to ensure that decisions are made based on all the available facts. This should be done in a transparent manner to allow all involved to understand the impact of changes. This is because a particular shape of building affects costs of building elements like foundations, walls, building structure frame, finishes decorations, roofing, electrical and mechanical services, which later also impact on costs of operating and maintaining the building and hence overall life cycle costs get affected. As such, a more accurate and still a simple number such as the CF should be quoted at these stages as it changes with design decisions.

In addition, because it accounts for conditioned volume, aspects such as floor to ceiling heights and internal layout of conditioned to non-conditioned areas will be picked up.

Layering in building design

Architect Frank Duffy CBE FRIBA uncovered the concept of "Shell, Services, Scenery and Sets" (layers) analysing buildings and their components in terms of layers of longevity. Identifying the layers of a building and their respective lifespan enables designers to consider the critical failure points of their design and accommodate for them. Stewart Brand extended these concepts in 1994 and the following layers are now commonly referred to in the building trade.

Site – lasts millennia and its residual value is linked closely to the level of pollutants and contamination found on it. The prospect of remediating contaminated land on which to build should be viewed through the prism of very long term returns on investment. And conversely the prospect of building on green sites should be assessed in terms of the negative effect it will have on the overall environment.

Structure – should last from a minimum of 30 years to over 300, the longer the structure is designed for the better the building. The longest lasting building across the world has excellent foundations and structural integrity.

Skin – designing for changing the exterior every 20–40 years can lend significantly longer lifespans to the overall building allowing aspects such as fashions and functional obsolescence to be accommodated without compromising the entire building. It also allows for less expensive adjustment for compliance with legislation for higher energy efficiency and airtightness as the climate and needs change.

Services – designing with clear understanding that the working and moving parts of the building need replacement every 7–25 years as they wear out is highly important and often falls through the cracks of the prioritisation process. However, there are many instances where buildings are abandoned or worse demolished because these short-lived systems are too deeply embedded to replace easily.

Space plan/internal layout – depending on the building type internal spaces may require re-arranging every 2–3 years especially for commercial buildings. In the past, many building types were designed with no real vision for future changes and this rigidity is what leads to devaluation. More recently, with societal and climate changes, the option to adapt the internal layout of domestic and non-domestic properties to changing circumstances and needs with little cost and disruption is a powerful selling point that is simply too good to miss. The easier it is to adjust the space plan of a building the longer it is likely to stay a valued space and continue to offer benefits to its occupants.

Stuff – furniture, fittings, fixtures, finishes, technology, appliances, lamps, etc. change at a staggering rate compared to the actual building. Designing in the flexibility (from siting enough power outlets in appropriate places to allowing for building wide network coverage and providing easy to replace wall and floor coverings, etc.) to allow for that change provides opportunities for adaptation and significant added value to occupants.

Understanding the interdependencies of these layers forms the cornerstone of flexible and enduring design.

Materials

The procurement of materials is arguably one of the highest costs for a construction project. The production use and disposal of materials is one of the highest contributors to global environmental problems as around 30% of industrial energy consumption is attributed to the manufacture and transportation of building materials. Furthermore, building and construction activities in general consume 3 billion tonnes of raw materials annually worldwide, which is 40% of total global use.

It should be noted that embodied energy is only ever a small part of a building's life cycle consumption (for a given 60+ year life span); nonetheless the choice of materials also affects its performance and thus influences its resource requirements over its life span. This is because the choice of materials and building elements affects thermal and structural properties.

Cost and environmentally effective material selection and procurement is paramount to developing an effective building. If done correctly, a building should have just the right amount of quality materials with minimum comparative environmental impact.

Local availability

Using locally sourced materials for the construction process helps to reduce delivery miles and directly affects impact. This can reduce carbon emissions for the construction company and is translated to embodied carbon for the building.

Overall local procurement impacts on aspects not immediately tangible but nonetheless important such as:

- reducing overall traffic on the highway network;
- shortened supply chains leading to cost savings;
- efficiencies with respect to construction schedules enabling just-in-time delivery allowing for less area on site for storage;
- reduced risk of damage to, or loss, of materials from prolonged storage;
- reducing associated risks from construction site congestion such as that of safety incidents;
- greater quality control as it is more feasible to travel to suppliers' sites to manage and develop the required products specifications;
- increasing investment in the local community providing corporate social responsibility benefits.

These, when taken in their entirety, permit comparatively significant cost savings.

The associated radius distance that is considered local does depend both on the material types procured, as it does on the location of the site itself and the constructors involved. For example, concrete procured from local batching plants should be prioritised even if re-enforced steel may have to come from further afield. Supplies such as major drainage, pipeline and waterproofing materials can be sourced from 5 to 40 miles of a project, and even though they do not make up a significant proportion of the overall project material value or volume they provide small efficiencies that can add up. This makes allowances for the relative scarcity of some "specialised" material types that are necessary for the building whilst allowing as many of the other supplies to pick up the slack.

It should be noted that the location of the project plays a significant role; for a site relatively close to a significant transport node such as a port or rail freight centre, the carbon impact of sourcing materials from a significant distance away will not be as high, but the risks associated with not getting the materials on site just in time will increase proportionally.

Finally, the transport and logistics infrastructure of the country, in which the project is based, greatly affects the effective transport radius to be considered. For example, a material can be considered local if it's within a 30 mile radius in a small country with an average transport network, whilst it can be 250 or 500[1] miles for a country such as the USA.

Identifying the types of materials plants and suppliers that are effectively nearby should be considered in the decisions at early stages (inception and concept design stages) by architects and structural engineers this will allow the constructor to provide significantly more competitive pricing and timing estimations when it comes to tendering.

There are a few aspects to balance against the benefits mentioned thus far with respect to local sourcing. For example, a number of projects (primarily procured for the public sector) have contract clauses requiring adherence to strict policy on competition; thus positive discrimination of local suppliers where a close relationship between supplier and constructor may not comply. Additionally, there are risks associated with the use of smaller business where the economies of scale reduce efficiencies for the process. Finally, the risk of the small business becoming too dependent on the buyer for the majority of its trade needs to be acknowledged and reduced, whereby the hazard of the project being terminated with significant local connotations or of the supplier becoming complacent and reducing quality should be addressed.

Complexity

Often the complexity of a specification affects the cost of the construction both in terms of cost of supplies as well as effort and cost to assemble/construct. This is only the first direct correlation.

The type of material, chosen at concept design, will influence the workmanship required at construction, and this affects quality control on site. The more complex a material is to specify, the more care is needed on site, and the more is required from the contractor with respect to taking cognisance of additional material such as quantities, schedules and drawings whilst constructing. This often also increases time to design and time on site because of the relative skills shortage associated with more complex systems.

A product such as concrete needs a moderate level of care when specifying, as it is a common material used the world over across all types of construction. As such the skills needed to specify a good product that lasts through years with little risk of defects are well established across the industry and can be found easily without special consideration given to the choice of designers/specifiers/procurement professionals needed for the task. Detailing, mixing and casting are the core variables to be aware of and again the worldwide skill base for ensuring good workmanship is abundant.

Conversely, there are more finicky materials used more commonly on site that require skill at specifying as well as detailing to ensure consistent performance without problems otherwise known as unforeseen costs, for example, composite materials. These are not only relatively new, meaning the skills are simply not there yet in the market, they are also relatively more expensive, but their use is increasing because they present other benefits such as increased strength or durability with lower weight or thickness. With these types of materials great skill is needed in specification to ensure the right materials are used appropriately and at the correct quantities. Additionally when detailing these materials, skills are needed to ensure they are adequately combined and incorporated into the building.

Beyond just the skills needed by the designers at early stages of building procurement, once the building has been tendered specification complexity impacts on effective value engineering. When the instance arises that the project requires changes so that the contractor can allow for programme or availability an overly complex specification can lead to significant failures in substitutions. The more complex a system the greater the difficulty in finding or approving "equal" replacements. Thus, substitutions arise that fall short of the required performance and sometimes this can impact on basic building needs such as fire or earthquake protection.

Relatively simpler specification and detailing requirements allow for economies of scale on site through eliminating redundancies from duplicated activities; harmonising specifications for raw materials across similar product groups; volume consolidation across similar raw materials to allow for sourcing from a rationalised set of suppliers; and, standardising on site processes such as quality inspections.

Reducing complexity in specification, and in building procurement as a whole, is not an obvious quick win; however, it delivers savings from between 10% and 18% across the board consistently and it starts with understanding material choices. Simplifying material choices and specifications reduces costs by allowing raw material management; increasing supplier efficiencies; providing a better negotiating platform for material procurement managers; and reducing risks associated from a small skills base.

Managing complexity requires effort in terms of planning, appropriate time investment and allocation (this does not mean a longer process simply a different process than the norm) and buy-in from all stakeholders, but most importantly by those that are further removed from the design and procurement side of buildings especially in the current silo mentality. A reasonable rule of thumb in this instance would be to consider that when specification and detailing are complex costs and the impacts of mistakes increase. Thus, a question that should be asked at regular intervals within the process is: "How much care in specification and detailing does this choice entail?" if the answer is anything other than an honest "reasonable", efforts should be made to simplify.

Durability and lifespan

The selection and specification of robust, durable materials in building design is the difference between a building that needs to be demolished within the designers' or owner's lifetime and one that lasts centuries. It is the difference between a money pit and true asset.

Buildings should be designed, constructed, operated and maintained so that, under foreseeable environmental and climatic conditions, they perform as required in their stated lifespan, with the necessary consistency for the safety of their occupants and the intended use(s) of the structure.

Durability is most apt for more modern light weight, easy-up constructions and it is important both with respect to individual materials as well as individual building systems. It should be noted however that, technically, there are no universally durable materials as this depends on the location specific climatic conditions and interactions with other materials in each application.

It stands to reason that the more durable a building, the more effective it becomes, the more economically viable it is for the owner and the more environmentally efficient. With respect to resource efficiency the longer a building is built to last the smaller impact it has on resources over a comparable building. And yet, standard costing and appraisal methods use lifespans of as little as 30 years; these timeframes are simply inappropriate when assessing a building. The most effective and celebrated buildings in the world last millennia and continue to provide revenue throughout their long lives both in economic and in social, human terms.

Longer lasting materials may incur a cost premium, and this is in terms of procurement as well as time on site; however, they balance this with aspects such as lower component replacement intervals, fewer maintenance activities and costs as well as thermal or energy efficiency.

In essence, more durable buildings are also significantly cheaper to insure and are intrinsically more adaptable to use and climate change. Robust design is all about managing the risks associated with the unique site conditions skilfully. The main aspects[2] to consider when assessing a particular material's robustness in these terms are addressed in the following text.

SERVICE LIFE AND EXPECTED WEAR

This is the period in which the building component is expected to function usually from 5 to 100 years. It is also dependent on accessibility and the ease with which the component can be repaired or replaced. This is where detailing also comes into play whereby components that need regular maintenance should not be inaccessible or so integrated with longer life components that they can't be repaired or replaced.

For inherently longer life components wear plays a significant role whereby excessive wear (such as more hostile environments) will increase maintenance requirements and/or reduce life span.

It is important to assess all individual components and systems in a building and design for easy maintenance. The main aspect to consider is the economic life of an item. The shorter the economic life, the more accessible for maintenance and repair it should be. For example, a typical lamp will last approximately 3–7 years whereas the building structure will be expected to last over 100 years. When integrating materials and components account should be taken to allow for these discrepancies, for example, if using concrete, access should be designed in for pipes and wiring that will need replacement at least three times in the life of the building.

If this is not detailed appropriately, the damage done by the shorter-lived components will have significant cost and time implications for the building, and, in the worst cases, may lead to the entire building needing to be demolished.

STRUCTURAL LOADS

Choice of materials should be based on safety first with considerations such as stress, strain, deflection and differential movement being paramount. The role of the structural engineer on the project is to ensure that the materials chosen will not fail during the lifetime of the building. In many cases short term profit priorities lead designers to reduce the expected lifespan of the building from say 100 years to 60 or from 60 to 30 to accommodate cheaper materials. This puts pressure on designers to provide specifications with less durable materials on the technicality that the expected life of the building is lower than that of its material components. Often the connotations of this practice are not discussed in detail with all stakeholders and, even though this is a fully legal and accepted process, it is very short sighted and simply not sustainable.

MOISTURE AND WATER MANAGEMENT

Moisture can affect a material in many ways such as corrosion, freeze-thaw spalling, mould or rotting which depend on the time and severity of expected exposure. Beyond the structural strain moisture puts on the building, other important components are also affected such as insulation.

Interestingly, technical data sheets produced by insulation manufacturers historically did not report fully on the performance impact moisture has, they stated hygroscopicity but not the, often significant, reduction in thermal properties associated with wet insulation. Add to that the fact that insulation is notoriously inaccessible for maintenance or replacement, there are many buildings that have been insulated to fully comply with or exceed building regulations that perform abysmally for decades due to moisture. In more recent years, some more progressive manufacturers, forced by consumers, have started reporting on the de-rating effect of excessive moisture. The practice is not yet fully widespread and so this aspect falls on designers and intelligent clients to address.

HEAT AND THERMAL STRESS RESISTANCE

Structural components and/or mechanical elements, such as pressure vessels and pipes, are subjected to thermal loads due to high temperature, high temperature gradient and cyclical changes of temperature. These stresses affect the performance and the life of the material if they are not appropriately addressed. Although data is widely available for the thermal performance of materials, the effect of climate change and the potential for the building to adapt to new extremes needs to be actively addressed for a building.

SUNLIGHT

The potential degradation from ultraviolet light is a significant factor that leads to materials being replaced at shorter intervals. UV light affects most plastics, roofing elements, wood, many paints and fabric building elements. UV stabilisers can be used of course, but they can prove environmentally harmful in other ways (such as toxicity). The potential for more frequent replacements should be a factor when considering the balance of material choice and capital cost for a building.

ATMOSPHERIC POLLUTANTS AND OTHER ENVIRONMENTAL FACTORS

There are many pollutants that affect materials; the main ones are sulphur dioxide and sulphates, nitrogen oxides and nitrates, chlorides, carbon dioxide and ozone depending on location marine environments also cause significant degradation. In city centres particulate matter, especially from diesel vehicle emissions, is of increasing significance. Damage can be in the form of loss of mass, changes in porosity, discoloration and embrittlement.

In terms of the building structure, fabric and systems, the materials most sensitive to pollutants are calcareous building stones, metals especially ferrous, polymeric materials such as rubbers and paints. A number of national and international programmes have been established to assess the responses of a range of materials to different environments. This data is used in numerous methodologies developed to allow estimates of the costs of damage from these factors and is available in many countries through environmental agencies and internationally through the United Nations economic commissions such as the Economic Commission for Europe (UNECE). These methodologies and data are well-established and are continually being refined with better data on rates of decay.

This information should be used to assess the action needed to avoid or reduce degradation of sensitive materials and buildings. This is of significant importance when considering overall building durability and whole life value. It is also essential when considering the potential effects of climatic changes.

The most recent estimates by the Department for Environment Food and Rural Affairs (DEFRA) in the UK calculate the cost of damage to rubber materials to be around £85 million per year.

Beyond just the impact on the building fabric and its systems pollutants can cause significant degradation to occupants as well as items within buildings such as books and artefacts. These pollutants may either be lower concentrations of pollutants originating outdoors or other substances generated from synthetic materials, paints, varnishes, etc.

BIOLOGICAL LIFE

The exposure of materials to microorganisms, such as bacteria, fungi and algae and macro-organisms such as insects should be considered during design and specification. Damage can be in the form of mineralogical, chemical and structural damage to the material (biodeterioration) or it can be visual through coloured biological stains on façades and roofs or it can affect the quality of indoor air.

It is ultimately cheaper to design so as to avoid these risks altogether if possible, for example, to avoid the use of wood for inaccessible building components where termites or beetles are known to be an issue, and where necessary to use specially treated wood.

OCCUPANT AND VISITOR USE

Where durability and robustness cannot be designed in by material choice alone exposed parts of the building internally and externally should be protected to negate the need for replacement more frequently than necessary. Serious consideration should be given to areas such as entrances; corridors, lifts, stairs, etc.; internal vehicle and trolley routes in storage and delivery; kitchen areas; and vehicle delivery, turning and access areas close to the external building fabric. It is considerably less expensive and easier to replace components such as bollards or to maintain raised kerbs than to replace damaged external walls. Similarly, it is simpler to maintain kickplates or some other form of impact protection than to replace the entire door. Aesthetically pleasing, hard wearing and easily washable wall and floor finishes could pay the marginal capital cost uplift off within the first couple of years of occupation when appropriately detailed.

For some building projects, stakeholders consider even these simple solutions as too much of a capital expense and rely on behavioural policies to maintain low construction costs, such as restricting occupant and visitor access to some areas to reduce the risk of damage rather than designing in a better solution that includes for the human element in building use. This again is relatively common practice when immediate profits are in any way threatened; however, it is a clear indication of simply bad design, short sightedness and complete disassociation from the real reason buildings exist.

Flexibility

A material's flexibility is an important component in design. Many materials can only be used in certain ways with significant restrains such as composites, insulations, glass, etc. Others, like timber based products, are extremely flexible whereby they can be used for anything from structure to finishes. The more flexible a material, the more freedom designers have in its deployment around the building thc casier it is to create a building able to adapt to external changes over time.

Maintenance and repair

Building maintenance represents over 5% of a country's annual Gross Domestic Product. It represents 45% of the total turnover annually in the construction industry. The costs of maintenance are hugely significant, and yet they are often underestimated or simply disregarded in the early design processes. This is particularly true for many public bodies and multi-building estates whereby the budgets for maintenance are separate to those of capital project investment. Also, it is extremely common among building developers looking to develop building and sell them on as quickly as possible, to simply not report on maintainability.

With maintenance costing as much as five times the capital/acquisition cost of a project over its lifetime disregarding it, is simply a bad business decision.

Badly maintained buildings incur higher insurance premiums and can also become uninsurable leading to even higher running costs. There are even cases where buildings, especially those visited by the public, have been deemed uninhabitable because of lack of insurance.

Clients and building users are only recently realising that maintainable buildings should be a priority, beyond aesthetics and capital cost. The rise in calls for "soft landings", "low running cost" buildings and whole life costing with accurate maintenance costs is only about 10–20 years old, although the realisation that running costs are directly linked to early design stages started being recorded in the 1930s.

Obtaining data to benchmark maintenance costs for building systems is easy for professionals within the design team. All professional institutions and especially those for cost professionals produce data annually on component, system and building wide maintenance requirements, practices and costs, with some professional bodies such as the Royal Institution of Chartered Surveyors (RICS) in the UK producing guides for building users as well.

Maintenance is necessary to preserve assets and protect building occupants. It also helps keep the value of the building higher for longer and, where appropriate, extending its income making potential.

On the flip side, it is a fallacy to choose "maintenance free" materials and systems that are not well thought through. In many cases the term does not imply fit and forget, rather replace altogether when broken which contradicts the principles of designing for durability and long life.

Environmental impacts

There is a multitude of environmental impacts associated with material choices to consider. Impacts occur across the life cycle of each material in question.

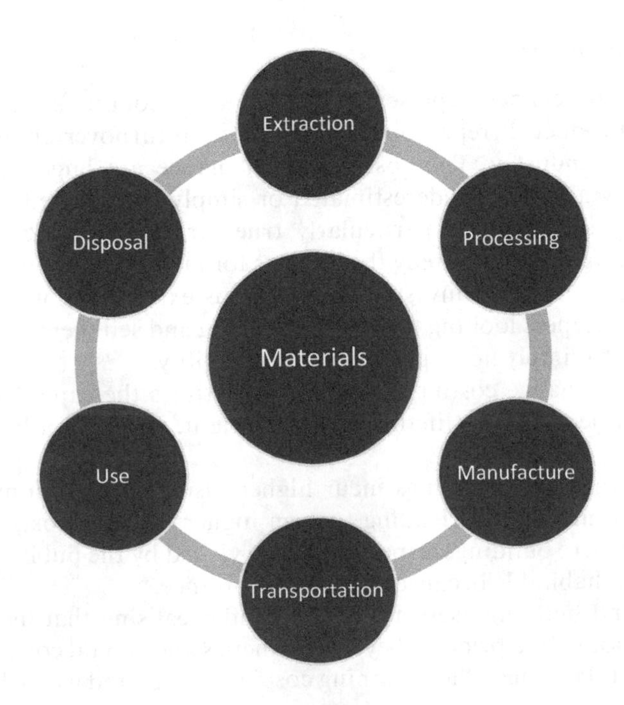

These can be categorised and include:

- Global warming potential (GWP) or greenhouse gasses. This is calculated based on the relative global warming effect of the material after 100 years of simultaneous emission of the same mass of CO_2;
- Water extraction based on the effects of depletion, disruption or pollution of aquifers or rivers and their ecosystems due to over extraction;
- Mineral resource extraction implications applying to all minerals including metal ore. Although the minerals may not be considered scarce the local environmental impacts from their extraction can be considerable;
- Stratospheric ozone depletion potential (ODP) calculated to estimate the damage caused by the increase to harmful ultraviolet light reaching the earths' surface;
- Human toxicity to air and water assesses the potential toxic effects of the emissions of substances such as heavy metals have on human health;
- Ecotoxicity to water and land assesses the potential toxic effects to the ecosystem beyond just humans;
- Fossil fuel depletion with respect to coal, oil, or gas consumption;
- Eutrophication (nitrates/phosphates) impacts assess the effects of increased concentrations of useful chemical in water, such as phosphates and nitrates, which can lead to excessive growth with negative impacts on oxygen production;
- Photochemical ozone creation at low level in the atmosphere leading to summer smog which impacts on human and plant health;
- Acid deposition assessing the ecosystem impairment by acids such as ammonia, hydrochloric acid, hydrogen fluoride, nitrous oxides and sulphur oxides;
- Habitat degradation measuring the impacts on ecosystems from habitat destruction, depletion and change;
- Off gassing measures the effects that the release of volatile organic compounds (VOCs) and other harmful chemicals has on the air quality around their immediate environment;
- Waste disposal measures the effect on landfill capacity along with the environmental effects of landfilling itself such as gaseous emissions and leachate pollution as a result of the manufacturing process of the material.

It is a complex process to understand the full impact of materials and the quality of data is a mixed bag depending on local trends and general opinion fashions. It is also difficult to choose the level of detail to go into for each category. Thankfully, more so than for other aspects of this process, there are considerable resources and tools available to help.

ECO-LABELLING AND CERTIFICATION

There is a sea of eco-labelling and certification schemes across the world. There are as many long established labels offering decades of market presence as there are new ones on the market for a couple of years. For manufacturers they make good business sense as they offer a potential competitive advantage. As such, they are easily available for most major products such as aggregates; bricks and clay; cement, concrete and blocks; wood; metals; and insulation among others. In general, the proliferation of such schemes has helped raise overall environmental performance across the board. The International Standards Organisation (ISO) stipulates that the act of communicating verifiable and accurate information, that is not misleading, on environmental aspects of products and services, encourages the demand for and supply of those products and services that cause less stress on the environment, thereby stimulating the potential for market-driven continuous environmental improvement.

However, this trend also poses a problem whereby manufacturers are tempted to make exaggerated or misleading claims, which confuse procurement professionals into thinking products are better than they really are. Instead of raising standards, the result is confusion and a systematic undermining of all eco-labelled and certified products. It is thus important to choose the scheme for each material certificate wisely. The label should be credible, independently verified and product specific where appropriate.

Timber certification is the most obvious and the easiest to source with minimal additional cost implications; the most popular schemes include but are not limited to:

- The Forest Stewardship Council (FSC) which certifies forest management and chain of custody and license retailers to promote FSC labelled products;
- The Programme for Endorsement of Forest Certification (PEFC) is a mutual recognition scheme which focuses on sustainable forest management and on the processing of timber, placing emphasis on the supply chain;
- The Canadian Standards Association (CSA) sustainable forest management scheme is applicable in Canada and is a 3rd party certification scheme endorsed by the PEFC;
- The Malaysian Timber Certification Council (MTCC) operates the Malaysian timber certification scheme which is a voluntary national scheme for independent assessment of forest management practices and audit of timber product manufacturers or exporters. It is also endorsed by the PEFC;
- The Sustainable Forestry Initiative (SFI) operates a number of standards that promote sustainable forest management in North America and is responsible for procurement of forest products around the world;

- The Earthworm foundation (formerly the Forest Trust (TFT) and the Tropical Forest Trust) is a global environmental charity that helps companies run responsible supply chains predominately with respect to palm oil, pulp and paper and timber to allow traceability;
- SmartSource (formerly Smartwood) is a Rainforest Alliance framework that establishes a legal, traceable and sustainable supply chain for forest products.

A main rule of thumb to ensure a level of credibility for any label sought for a material, especially those that look "new", is that the organisation issuing these labels is accredited under ISO 14024, *Environmental labels and declarations – Type I environmental labelling – Principles and procedures*. The standard establishes the principles and procedures for developing environmental labelling programmes, including the selection of product categories, product environmental criteria and product function characteristics, and for assessing and demonstrating compliance. It also establishes the certification procedures for awarding the label thus defining the competence of verifiers and providing the rigour and helping make the scope clear reducing the possibility of misleading labels and "greenwashing".

Building materials such as steel, concrete and masonry do not provide as much labelled certification choice yet. Nonetheless, there are schemes in place that can be used by procurement professionals throughout the building project to provide credible environmental credentials. These are mainly in the form of:

- approved environmental product declarations (EPDs) directly from manufacturers;
- approved environmental profiles provided by independent organisations through numerous[3] online databases;
- full Life Cycle Assessments of on an ad hoc basis covering an individual material or the entire building as appropriate.

ENVIRONMENTAL PRODUCT DECLARATIONS

The Green Guide to Specification is part of BREEAM. It contains more than 1,500 specifications used in various types of building. It is a live resource that reflects changes in manufacturing practices, the way materials are used in buildings, and evolving environmental data;

GreenSpec is a "Green Building" resource in the UK. It is an independent body that identifies and endorses green building products, systems and services using the PASS (Product Assessment Screening System);

Ecospecifier Global specialises in categorisation of products into Building, Hospitality, Health & Beauty, Personal Care and Cleaning Products and complies with the requirements for schemes such as Green Star and LEED. It provides region specific information for Australia and South Africa;

Good Environmental Choice Australia (GECA) is Australia's independent ecolabelling program and provides a comprehensive database for building products and materials;

The international EPD system database (EnvironDEC) provides environmental declarations based on ISO 14025 and EN 15804 for a wide range of product categories by organisations in 43 countries.

With respect to construction products in particular, it is a good idea to double check that the declaration is compliant with one or more of the following as applicable:

- ISO 21930 Sustainability in buildings and civil engineering works – Core rules for environmental product declarations of construction products and services;
- EN 15804 Environmental Product Declarations (European standard);
- ISO 14044 Environmental management – Life cycle assessment – Requirements and guidelines.

The use of labelled or certified products simplifies the complexity of assessing the environmental impacts of products and, even if a whole building certification is not sought, can go a long way into reducing the negative effects of building construction. As more and more manufacturers and suppliers are acquiring these certificates for their products the risk of uplifted cost associated with a certified product reduces considerably and so this exercise can be cost neutral if considered appropriately at the beginning of the design process and monitored throughout the project delivery phases.

LIFE CYCLE ASSESSMENTS

The LCA process is complex because it is comprehensive and it evaluates the environmental impacts of a system taking into account its full life cycle, from cradle to grave. It considers the impacts associated with the production and use of the system. There are many tools available to carry out an LCA and as such it can be difficult to choose which tool to use. Any tool based on the methodology set out in the ISO 14040-14044 series (life cycle assessment) is reliably consistent. Ultimately the quality of the output from such an assessment is directly correlated with the quality of the input data and the methodology used to analyse it.

The most common database for LCA data is ecoinvent. It provides process data for thousands of products that feeds directly into many of the software packages available for use without special license or conversion. Information can be found from many other databases such as the US life cycle inventory database, the life cycle data network (LCDN) in Europe and the Japanese Inventory Database for Environmental Analysis (IDEA). With increased use of these tools the demand for transparent, reliable and consistent data has meant that in the past few years information of this sort is much more readily available than it has been making its use easier.

Common systems/software available to use for selection of materials/constructions at the early design phase include but are not limited to the ones listed in the following section. These tools contain LCA data sets for generalised materials and constructions, and the results are "multi-dimensional" in the sense that they do not rate the environmental performance of the building, such as the tools described in previous chapters, rather they provide a comparison base for different design decisions. Resources used in these tools have been developed by both the public and private sectors. It is advisable to select a tool appropriate for each project as different tools offer different functionality, complexity and results.

In deciding to carry out an LCA an investment in time and money is needed by the project team and stakeholders. Additionally, the persons carrying out the assessment need a level of specialised knowledge to ensure interpretation of the results is appropriate and the reporting of the findings is useful. It is not a tool to be used in isolation, rather as part of additional actions by the stakeholders whereby the scope of the LCA analysis is well understood. To maximise the gain of this investment the assessment should be revisited at different points in the building procurement process to evaluate the impact of changes (sensitivity analyses).

The Integrated Material Profile and Costing Tool **IMPACT** (which now incorporates what used to be ENVEST) is the life cycle environmental impact assessment design tool from the BRE for use through the earliest phases of commercial/mixed use building design. It provides information for both the operational impacts and the materials impacts of a building as the design evolves. As a tool it is helpful in that the user can identify trade-offs to minimise greenhouse gas emissions and other impacts over the life of the building. It also provides cost information for these decisions on a capital investment and whole life level. It can be used on a whole building as well as an individual material level. It is a tool that can be incorporated into existing systems such as whole building energy simulators.

SimaPro was developed by PRe Sustainability and is one of the most used licenced LCA software package with a 25-year market presence across 80 countries. It can be used for a variety of applications, such as sustainability reporting, carbon and water footprinting, product design, generating environmental product declarations and determining key performance indicators. This tool uses a combination of many databases from across the world and as such provides a more comprehensive assessment.

GaBi is an equally comprehensive licenced LCA software package with the additional option of being able to ask the Thinkstep developers to provide information for gaps in the databases. This tool also provides graphic outputs to help illustrate results in useful ways to allow decision-making.

The **Athena** Sustainable Materials Institute in Canada has developed a suite of free software tools to help conduct LCAs. The Athena EcoCalculator tools are spreadsheets with pre-defined assembly and envelope configurations for quick but limited estimates of building design footprints. The Impact Estimator for Buildings is a stand-alone program that allows more detailed modelling of specific designs and allows for existing building applications and includes for the input of energy simulation results to calculate operational as well as embodied effects. Pavement LCA (formerly Impact Estimator for Highways) is an LCA-based software tool that measures environmental impact of roadway designs. The benefit of these tools is that although users do need some technical knowledge, they do not have to specialise in the field to get good comparative results.

The Building for Environmental and Economic Sustainability (**BEES**) software was developed by the US National Institute of Standards and Technology (NIST) Engineering Laboratory. It is an online tool and includes data for limited amount of building products. However, it is one of the few systems that assesses performance based on the American Society for Testing and Materials (ASTM) standard life-cycle cost method, which covers the costs of initial investment, replacement, operation, maintenance and repair and disposal using Multi-Attribute Decision Analysis.

Prioritisation

It is important to avoid getting too bogged down with these elements when considering the entire build project. Some level of prioritisation is recommended based on the details of the building. Materiality assessments carried out by certification organisations over a period of years have indicated that some elements have a bigger impact than others.

Importance should be placed on aspects such as the external thermal envelope (roof, walls, floor and foundations) the internal floor slabs, coverings and internal walls and finishes and partitions, external windows and doors.

It is important to set overall policies for material types from the onset of the design to ensure all parties are working towards the same goal. Identifying materials by general category can help with this process. For example, breaking elements up into groups such as:

- Metals (steel, aluminium, etc.);
- Concrete (including blocks, tiles, etc.);
- Clay (including bricks, tiles, etc.);
- Stone;
- Glass;
- Composites;
- Timber;

- Plastics;
- Additives;
- Insulation;
- Adhesives.

Comprehensive procurement policies, taking this section's considerations into account, can then help designers with their specification and detailing decisions and contractors with their procurement and construction practices. It is important also to explain the reasoning behind the policies to finance and cost orientated stakeholders to avoid the risk of dilution of these principles as the project progresses.

Waste

Wastage is expensive, much more expensive than actually acknowledged, equally in terms of money, of environmental impact and with respect to time on site and during building use. The environmental cost of waste is undeniable.

As a rule of thumb waste uplift costs are typically less than 20% of the true costs of wasted raw materials. The majority of this cost includes procurement, handling and processing which are generally not factored in. So, for example, in the UK, for an average skip full of construction waste only 15% of the cost is associated with the actual skip hire and disposal, 76% is associated with the cost of the materials in the skip and the rest is associated with the cost of labour such as handling. Reducing raw material waste through better design, storage on site, handling and manufacturing procedures will thus proportionally reduce construction time, procurement and waste costs.

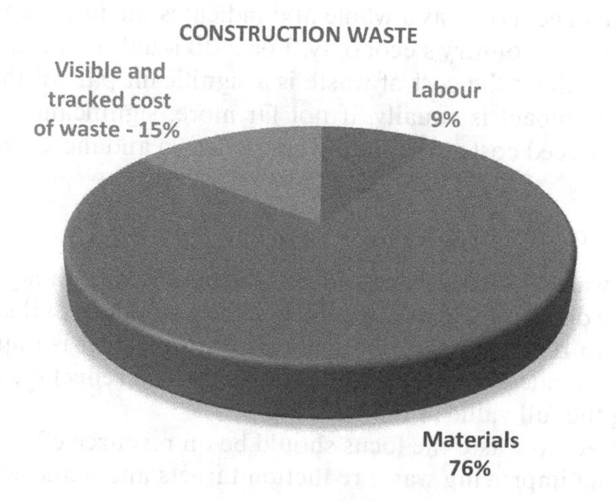

During the typical construction project, it is accepted without much question that the main construction contractor will abide by local legislation or environmental certification requirements and develop a basic waste management plan as part of the project delivery agreements. In the UK this is required by law. In most cases some form of records are exchanged as an indication of compliance. The reality is that construction organisations, save a very small minority, do not place a large enough emphasis on full effective waste management and there is considerable room for improvement across the board.

Waste monitoring is now considered common practice. Although monitoring allows an organisation to identify the areas where the most significant improvements can be made, analysis of the data collected sometimes falls short. Actions are not always effectively completed once a trend is identified (if it is identified) with many contented at collecting the data to tick a particular box wasting the effort taken to collect the information and ultimately losing out on the potential for savings.

Other more progressive organisations may set targets and objectives on an organisational level or and/on a project by project basis. Again, many of these organisations miss a trick; the act of setting targets tends to satisfy most of the legislative requirements out there with respect to everyday building construction. Thus, again many are content with simple compliance and either set easy, token targets for a short spectrum of waste streams or fall short of creating robust and enforceable action plans to achieve more ambitious targets and objectives.

This fallacy affects capital costs for buildings across the spectrum of the process; in 2015 Zero Waste Scotland identified that, on average, setting and importantly achieving a target to reduce the overall quantity of all waste generated by 10% could save £6 per staff member across the private and public sector, per year (not including material purchase costs or other "hidden" costs of waste such as handling time). This translates to hundreds of £millions for the Scottish economy as a whole and indicates the importance of reducing waste for any country's economy. For individual organisations such as constructors, where the cost of waste is a significant part of the company's turnover, the impact is equally, if not far more, significant and translates directly to reduced costs for building construction and increased profits.

Waste reduction by design

The earlier waste is seriously considered the better. Well thought out design is core to the delivery resource and cost efficient building that lasts. It is also paramount to realising whole life value for a project. It is important to be clear about intended objectives and outcomes which reflect the commitment to realising the full value of the building.

With respect to waste the focus should be on resource efficient construction including improving waste reduction targets and achievements, use of materials with higher recycled content, re-use of materials, combined with consistent and effective monitoring, analysis, evaluation and reporting. The process starts with reducing the consumption of raw materials at the outset, optimising resource-efficient maintenance and repair and maximising re-use of materials at the end of life.

Traditionally, waste reduction in the building trade has been reactive responding to legislation and certification requirements and the financial pressures of rising landfill costs. As a result, processes and project actions which actively reduce waste by design are considered "new" and "innovative" which all too often also translates to "optional" for many finance professionals and Clients. To add to the problem, many companies offer a level of waste reduction as "free" added value which further devalues the efforts and makes them look even more optional. This is not a true reflection of the value of designing out waste.

In fact, the concept of waste hierarchy was introduced to the EU in through Directive 2008/98/EC on waste (Waste Framework Directive). It has since been introduced in legislation for all member states, similar directives are in place on countries around the world. As such the practices of minimising waste in building construction should not be considered optional and time should be allowed within the process to take necessary measures as needed.

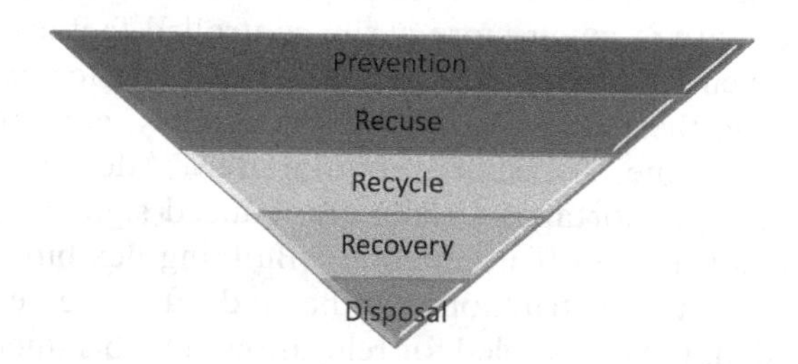

WASTE-EFFICIENT PROCUREMENT

The buy-in to waste reduction by all involved in the project is paramount. Early adoption of waste reduction options is required across the board, even if this means departing form "standard", "business as usual". As this is a relatively new process for the industry opportunities may arise to acquire materials or components beyond conventional procurement routes. Protocols, policies or exceptions need to be allowed for flexibility within the contract for the project.

The basic starting point is material optimisation. The client, design team and subsequently the contractor will need to invest time on the most efficient use of resources by favouring design solutions that lead to a significant reduction in waste generated both during material manufacture and during construction and building use. Examples of this include minimising excavation, using intelligent cut-and-fill methods to minimise waste generation and the need for virgin materials, simplifying and standardising materials and components, as well as creating an efficient building shape through dimensional coordination. Taking the principles of this book into account and designing for sequential access to relevant services to prevent material damage during maintenance or repair is also a proven method of reducing waste. Using durable materials also provides benefits with respect to waste during the operation of the building. Finally, simply minimising and standardising the type number and type of fixings and components used when detailing makes a considerable impact during the building's life when done appropriately.

Volumetric constructions, sub-assemblies, components, pods and panelised systems manufactured off site along with site-based manufactured methods of modern construction (MMC) can contribute to overall waste reduction of over 20% compared to projects using exclusively traditional construction methods with some individual solutions providing savings of up to 90% with respect to on site waste. This presents considerable improvement and provides many additional benefits such as higher build quality resulting in better building performance and quicker build times paying back for the potential increased costs within the duration of the construction project.

Ultimately the practice affords increased construction speed as indoor production lines are not affected by adverse site conditions such as the weather or ecological constrains associated with timing for flora and fauna activities. As quality is much more easily controlled factory environment building component characteristics such as acoustic performance, durability, fire resistance, thermal performance, consistency, etc. are more likely to be correct first time if specified appropriately. Additionally, the risks associated with skill shortages on site are reduced significantly for parts of the construction using offsite methods. Building flexibility can also be increased as modular constructions can be (and often are designed by default) assembled and disassembled for relocation, refurbishment and re-use. Finally, considerable benefits can be realised with respect to the environmental intrusiveness of a building site with respect to dust, noise and overall disruption. It should be noted that offsite construction does require some changes to the standard, traditional on-site practices and therefore, it needs to be specified early in the design process.

Reuse

Prioritising the application of re-used or recovered materials may also be an option, depending on project particulars, by considering the possibility of re-using existing structures on site, sourcing reclaimed products, or using crushed demolition materials wherever practical and safe. It is worth investing a fraction of additional time during the consultation and design process to assess the economic and environmental benefits of using reclaimed materials.

The use of reclaimed materials is not as risky as some may consider; demolition companies derive considerable revenue from the reclamation and resale of construction materials. The main cost increases occur because of problems with storage and double handling between sites. The ideal use of reclaimed materials is either on the same site, or one very nearby, to avoid excessive transport costs. So the aspects to consider include the distance travelled for the components as well as compliance with safety and quality standards.

Common components that can be re-used from existing building on site or from nearby with little or no re-conditioning include:

- stairs;
- timber such as hardwood flooring, weatherboards, laminated beams, truss joists, framing, posts skirtings, wood panelling, specialty wood fittings and joinery;
- thermal insulation such as fibreglass, wool and polyester insulation, polystyrene sheets; carpet and carpet tiles;
- plumbing fixtures such as baths, sinks, toilets, taps, service equipment, hot water heaters;
- electrical fittings such as light fittings, switches, thermostats;
- doors and windows such as metal and timber doors, mechanical closures, aluminium windows, steel windows, sealed glass units, unframed glass mirrors, store fronts, skylights, glass from windows and doors, timber and metal from frames;
- clay and concrete roof tiles;
- metal wall and roof claddings;
- PVC and metal spouting.

Taking these into consideration early and effectively may bring about changes to the entire build process. For example, a constructor that is appropriately engaged with these policies may make adjustments to their project plan to allow for on-site re-purposing of waste to enable on the spot reuse (e.g. crushing bricks, blocks and concrete from a demolition to create recycled aggregate for paths and roads). Stakeholders should be open to these minor changes to "standard" processes as the project progresses bearing in mind all the potential benefits of allowing for time to plan and implement waste reduction policies.

DESIGN FOR DECONSTRUCTION/DISASSEMBLY

Designing for deconstruction at the start of a project reduces in-use risk and cost by ensuring that building elements and products are designed to be quickly and easily maintained and replaced. It also retains a building's value beyond the end of its useful lifespan by making its components a resource for another building ensuring very efficient use of resources. Because of this, detailing for deconstruction makes a property a more attractive investment opportunity. In general terms, the process helps by increasing building flexibility and translates to real savings with respect to future costs for replacement, upgrades to comply with legislation, climate adaptation and change of use. It saves money in the short term with respect to incurring lower insurance premiums for latent defects insurance. And any potential increase in capital costs can be significantly minimised by engaged early consideration by stakeholders and open communication with constructors during the later phases of the project. Secondary benefits include the significant reduction of waste to landfill, helping the local economy, reducing transportation, reducing carbon impact and minimising pollution over the life of the building.

Design for flexibility of use and deconstruction, as well as climate adaptation, is a principle focusing on the whole life cycle of the building and is strongly linked to the design for re-use and recovery principle. The main difference between this principle and simple recycling is the idea of practical reuse, so it is a principle to be used alongside other considerations such as sustainable design and recycling.

There are many possibilities for re-use when proven viable alternatives to standard details are implemented, for example, the use of bolts instead of adhesives or the use of partitions to allow for internal space reconfiguration.

The principles are simple and include but are not limited to:

- Use of prefabricated, preassembled and modular elements that can be easily removed and replaced;
- Development of simplified and standardised connection details;
- Designing connections between construction elements so they can be successfully deconstructed;
- Using durable components to maximise the number of times they can be re-used;
- Designing the structure to allow for the greatest number of possible occupancies;
- Anticipating differential wear and tear enabling worn or unwanted surfaces, both internally and externally) to be removed without disruption elsewhere;
- Optimising opportunities for building services deconstruction through careful detailing and building layout;
- Simplifying and separating building systems such as distribution systems within non-structural walls to allow for selective removal of the low-value components.

Recycling

For any materials that cannot be immediately reused and that have to be discarded, a recycling policy should be agreed upon to minimise as much as possible the waste going to landfill.

Licenced specialist waste contractors have processes set up to maximise recycling. Clearly, as with all things, services vary; however, companies with proven high waste recovery rates should be preferred. It is important that materials are sorted on site as often as possible to provide better control and information for future works. Where space on site is limited some waste contractors can sort waste at their facility for recycling.

In the earlier project procurement stages processes should be in place to find a constructor with a clear sorting policy and well informed site staff. A site waste management plan that acknowledges the actions required by site staff should also provide details of how the main waste streams/types change as the build progresses.

There are numerous, readily available, guides to managing waste on site during construction, the main aspects of which provide tips on labelling, storage, compacting, separation and bin locations, staff training and sub-contractor management, as well as negotiations with suppliers to return packaging materials among others.

Circular economy

Circular economy thinking is at the heart of the thinking behind this book. On the whole, circular economy looks to maintain access to materials and resources for continual and future use and builds on themes relating to waste reduction, recycling, reuse, material efficiency, security of supply, sustainable consumption and production, better design, sharing of resources, etc. It is more than "just" recycling; looking at the entire life cycle of any process and considering how to arrive at the best whole life outcome. The concept looks to keep products, components and materials at their highest utility and value at all times through life extension and maintenance, reuse, refurbishment, remanufacture and finally recycling.

New design approaches are needed along with updated business models. The construction industry is one of the slowest to join the trend set across the globe embraced by profit earning entities such as:

- Ecovative design which supplies packaging and more to fortune 500 companies across the world;
- Biomason which grows biocement based construction materials;
- British Sugar that uses the waste products from sugar beet production to create saleable products beyond just sugar;
- ArcelorMittal in Brazil that has managed to sell or reuse in-house around 90% of what was previously categorised as waste from its flat steel production.

The adoption of circular economy principles reflects a growing global trend of forward-looking enterprises strategising their profit-making and investment to drive positive social and environmental impact. With the

Ellen MacArthur foundation spearheading the CE100 network of companies and governments creating tools and incentives to promote circular economy practices, there is significant incentive to adopt this approach on a building project with proven benefits shared across the life of the building.

Information and guides are readily available through organisations such as the European resource efficiency knowledge centre (EREK) and its supporters, the resource efficiency and sustainability task force and the environmental protection agency (EPA) in the USA, the UN international resource panel, the Australian chamber of commerce and industry (ACCI) and many more.

Location

There are significant interactions between the external environment, the building orientation and consequently the effects of the angle of the sun and the prevailing winds at different times of the year and the day, the building insulation and thermal mass the ventilation and heating systems and the casual and solar gains. These systems are interdependent and as such a change in one system will affect the balance of the entire thermal environment within the building. Thus, options adopted should be considered in relation to their possible effects on the entire building and by extension the building occupants.

The best method of assessing the thermal performance of the building is by dynamic building simulation. The simulation process involves the creation of a representative 3D model of the building under investigation, in realistic external environmental conditions (surrounding buildings and shading objects included, and a site specific weather profile), and analysing its thermal characteristics over time periods (1 day, 1 month, 1 year, etc.) and under changing climatic conditions using predicted future weather files.

With respect to business effectiveness, location is in many instances the most important aspect of building development. For organisations access and availability of the best and the brightest is one of the biggest factors in terms of location strategy and why companies might choose one location over another. According to a survey conducted by Cushman & Wakefield and CoreNet Global, cost is still the main driver for considering location decisions; however, talent availability, public transport and access to amenities and services are in the top 5 for North America, EMEA and Asia Pacific. Within the domestic sector, the higher property prices are achieved when aspects such as proximity to urban and city locations with quick access to amenities and easy access to facilities are part of the sales pitch.

Choosing to redevelop land that has previously been built upon or reclaiming sites significantly reduces the burden on precious (and often more expensive) undeveloped sites and greenfield space. Redevelopment also means valuable and effective regeneration of previously defunct sites such as those used for industrial or landfill purposes. These sorts of actions provide new life to ailing conurbation further enhancing communities whilst helping social as well as economic aspects of the area.

This book contains a non-exhaustive list of many of the individual aspects that could all be assessed to determine the most effective, low cost environmental, wellbeing and thermal balancing strategy for the building. It is strongly recommended that such analysis is carried out to assess the effectiveness of any initiatives/design options adopted for the building.

Orientation

Orientation has and, by extension, internal layout has significant effects on a building's performance with respect to its energy consumption during operation and the comfort it affords its occupants.

Orientation influences the solar gains for a building. Direct solar radiation increases the risk of overheating in the summer but is considered useful in the winter. Thus, choosing the optimum orientation to maximise daylight and to minimise summer heat gain and winter heat loss can have a significant impact on energy efficiency.

South facing buildings will face the daily path of the sun. This maximises daylight, but could create problems in relation to solar gains. Thus, some sort of external shading in summer is often required. East and west facing buildings are affected by variations in glare and solar gains at different times of the day, with instances of one side of the building overheating in the morning and the other in the afternoon, whilst north facing buildings are often the coolest and so are preferred in warmed climates where solar gains pose a threat to thermal comfort for the majority of the year.

Orientation also plays a major role with regard to prevailing winds. The layout of the site and the position of the building on the site will affect its level of exposure and protection from the elements and so care needs to be taken to ensure the lowest possible impact.

Depending on the site location, the use of trees, hedges or other barriers could be practicable. Orientation is very important in striking a balance and the following presents a non-exhaustive list of considerations when choosing the orientation of the building:

- Surrounding obstructions (trees, other buildings, etc.);
  - This is in conjunction with prevailing winds, the sun path, noise, pollution, etc.
- The sun path in relation to the building and the surroundings;
  - This will affect the direct solar radiation entering the property in different times of the year influencing thermal comfort and energy use to maintain comfortable internal conditions;
  - It will also determine the amount of natural daylight available, the risk of glare and will influence the energy consumption for lighting and the requirement for anti-glare devices such as blinds.

- The prevailing wind conditions;
  - This will influence the specification of openings on particular sides of the building to ensure minimal air infiltration;
  - Other considerations include details such as movable external shading devices which will need to be specified so as to minimise possible wind damage;
  - In essence, the building fabric and the building air tightness need to be such to minimise the impact of the wind.

Transport links/access

The quality and functionality of transport links and accessibility to a site directly affect a number of outcomes. Within the non-domestic sector business efficiency is directly correlated with transportation effectiveness.

Providing good transport links to a building frequently visited by the public increases footfall and has a positive impact on customer service. Within the education sector, transport links are paramount to, retain and recruit students whereas in the commercial sector good transport infrastructure is important in retaining staff.

On a more general level improving transport efficiencies is better for the environment and the community by helping reduce the ever present problems of increasing traffic congested roads. For central and local governmental buildings, it is seen as being vitally important to set a good example to the communities that they serve.

When designing a building there should be some thought placed at the early stages (such as site acquisition or feasibility stage) on how the occupants will get to the building. By planning ahead many initiatives can be incorporated with zero cost. Considering the options that will be available for commuters to access the building can promote the use of alternative modes of transport. The basic principles of sustainable transport planning and design focus on a pedestrian first approach.

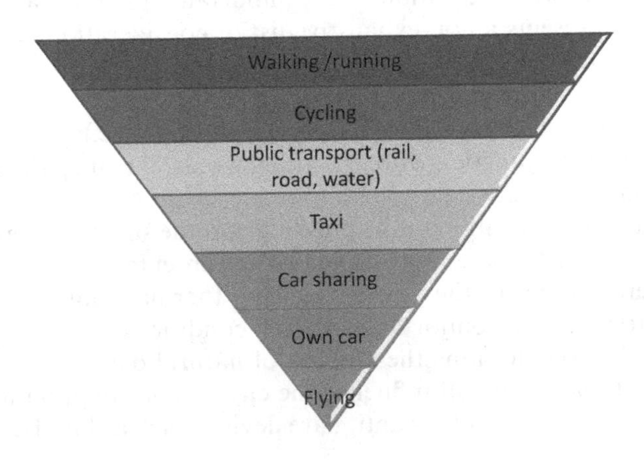

Designers and developers, in particular, have an enormous part to play in giving visitors and occupants access to good transport infrastructure. Developments within easy access to integrated transport systems should be prioritised thus reducing the need for cars. The culture of the main project stakeholders is paramount in these situations.

Looking at the bigger picture many governments are already heavily taxing certain types of cars providing incentives for many to simply opt out of car ownership. In addition, there are plans in place to ban diesel cars in many cities in the relative short terms with further plans to ban petrol cars in the next 20 years. Many city-wide plans are centred on encouraging the development of new travel patterns for their inhabitants based on the sustainable travel hierarchy as adopted by governments across the world. The financial incentives range from reduced governmental taxation/penalties to significant return on investment over the life of the building.

Optimising a building/development's location and minimising or negating the need for car parking provision or car parking spaces, complemented by alternative modes of transport to encourage the use of bus, rail and cycling, allows more of the site to be utilised. It can also mean that a smaller site can be purchased for erecting the building significantly reducing capital costs or that costs associated with creating vast and often multi-storey underground car parks are eliminated. A well accessible building also better retains its re-sale value.

Finally, driving less and accommodating for the reduction in car journeys provides considerable health benefits. Sustainable transport leads to fewer breathing and heart problems; it helps increase physical activity; it reduces traffic related injuries; it improves quality of life and reduces stress. Some facts:

- Vehicle emissions are a significant contributor to premature deaths and hospitalisations attributable each year to air pollution in any given city;
- The World Health Organization (WHO) reports that the health risks from air pollution can be up to 50% greater for people who live or work near busy traffic corridors compared with lower traffic areas;
- Exposure to air pollution from vehicles is linked to more asthma symptoms, reduced lung function, acute bronchitis, hospitalisation for respiratory and cardiac problems, cancer and reduced life expectancy;
- Reductions in traffic volume are linked to improved health. Reduced vehicle traffic in Atlanta Georgia during the 1996 summer Olympics was linked with fewer children requiring medical attention for asthma during this time;
- Areas, sites and neighbourhoods with low levels of road traffic, generous sidewalks, ample cycle paths and proximity to parks or playgrounds enable physical activity and, people who are physically active are less likely to experience heart disease, stroke, osteoporosis, colon cancer or diabetes than those who are inactive;

- The WHO reports that the most effective way to increase physical activity is to promote walking and cycling for shorter trips;
- Many sustainability and wellbeing certification schemes such as LEED or WELL provide incentives to increase walking and cycling opportunities for occupants;
- Traffic fatality rates tend to decline as public transport ridership increases in a community;
- The severity of traffic injuries is related to traffic speed. More pedestrians experience serious injury or death in higher speed zones than lower speed zones;
- The risk of injury to child pedestrians is related to traffic volume. More children are injured in high-traffic areas than lower traffic areas;
- Efficient public transport, cycle paths, safe walkways, ride-sharing systems and traffic calming measures improve the quality of life and the liveability of neighbourhoods;
- Traffic congestion creates stress for commuters, impairs work performance and can lead to aggressive driving behaviours that increase the likelihood of collisions;
- Traffic noise can lead to irritability, sleeplessness and depression;
- Communities dependent on cars (for example, many of today's suburbs) experience less opportunities and time for social interaction, leading to social isolation contributing to the increase in recent loneliness statistics;
- Transport links designed primarily around cars can limit opportunities for employment, access to services and social interaction among those who do not drive or cannot afford cars;
- Parking space occupies up to 15% of public land in sprawling metropolitan areas;
- On average, new roads become congested within 7 years and avoiding road travel altogether is more effective in reducing congestion.

The benefits of adopting, designing for and accommodating a sustainable transport design strategy include:

- Reduced capital costs from the reduction in car parking spaces needed;
- Reduction in site size needed and thus in land and resource use;
- Improved mental and physical health for occupants and visitors;
- Potential for easier planning permission where the development follows and enhances local and central government initiatives;
- Less risk from future legislative changes;
- Increased attraction to more of the population (not just car owners and users);
- Increased staff/student attraction and retention;
- Increased land availability for development and sustainable densification within urban environments;

- Improved air quality leading to potentially more options for servicing the building;
- Reduced noise pollution allowing more opportunity for natural ventilation.

WALKING AND CYCLING

Heavily trafficked roads present obvious barriers for those walking and cycling to a building; however, this can be overcome through the provision of safe and controlled crossing opportunities where pedestrian and cycle desire lines meet vehicle routes. Better yet, where there is the opportunity, crossings such as signalised junctions can be designed and programmed to prioritise pedestrians and cyclists.

Larger infrastructure cycle and pedestrian links can be designed in or connected to from the site. Paying attention to accessibility issues such as avoiding/removing steps and using dropped kerb crossings provide a suitable facility for cyclists, buggies/prams and wheelchairs and also prolong the life of the building and site by negating the need for later upgrades.

In many cases, additional costs can be saved by not providing roads for car access but instead providing off-road links between the building and established cycle routes or public transport access points. Appropriate lighting and surfacing are paramount to making these facilities inviting and effective as is the provision of pathways of a generous width for shared use. Facilities need also be provided such as cycle parking, wheelchair/mobility scooter charging points, pram parking and, if appropriate, changing, shower or clothes drying areas within a building. These additional aspects with respect to space and potential capital cost uplift can easily be accommodated when compared to the savings possible from reducing the car parking provision, and in most commercial buildings, the increased footfall from visitors will provide additional profit opportunities for the entire life of the building.

PUBLIC TRANSPORT

During the site selection stage, the provision of good public transport facilities should be prioritised as it provides significant benefits to occupants and visitors as well as businesses and organisations dependent on foot traffic. Use of public transport networks for commuting is affected by a number of factors relating to convenience and predictability. These are often impossible to assess without an in-depth study at the beginning of the project. However, there are a number of rules of thumb that can be used to quickly assess the opportunities. Things to look out for:

- The distance to a public transport stop (Bus, tram, rail, boat, etc.) should be easily walkable in all weathers;

- The origin/destination of the public transport provision should be a local urban centre;
- The frequency of the service should be appropriate for the travel patterns expected by occupants and visitors.

In essence, the availability of a good and accessible service significantly improves the likelihood that people will use it for travel to and from the building.

OTHER VEHICLES

Road transport is responsible for over 17% of greenhouse gas emissions in Europe. In the UK, much like many other western societies, over 60% of road transport emissions are generated by car and taxi use, equating to almost 15% of the total annual emissions for the UK across all sectors.

This, of course, does not account for the other impacts of road travel such as the environmental harm caused by roads, or the detrimental effect car parking has on the usable space of a site.

It will take generations for our society to move fully away from current, embedded travel and transport patterns; however, there are many things that can be done to ensure that a building and site developed today is as effective with respect to accessibility and transport and travel as it will need to be in 20 or 40 years' time. Much like the simple measures recommended for waste reduction, the reduction of road traffic and its impacts is split into three main steps.

- Avoid or reduce the need for unnecessary trips;
  - For example, by developing sites within easy access to the necessary amenities such as eateries, shops, medical facilities, businesses and schools, etc.;
  - Additionally, facilities for high quality, private and undisturbed video conferencing can be incorporated to greatly reduce the need for business travel.

- Shift travel modes;
  - Develop on sites with adequate public transport;
  - For example, shift the priority from single occupancy vehicles to car sharing or from car sharing to Public transport, etc.;
  - Provide incentives such as increased parking tolls or, more generally, congestion charging, vehicle ownership taxation, road tolls and public transport subsidies;
  - Provide adequate information on the availability of and access to public transport;
  - Aspects such as car sharing can be accommodated and encouraged if parking was prioritised for disability access and restricted to exclude single user vehicles.

- Improve the overall efficiency of each mode.
  - For example, by prioritising for high efficiency electric vehicles where they are necessary, such as buses, vans and cars.

Landscape, ecology and flooding

The majority of projects addressed by this book will either be in an urban or suburban environment or be the precursors for the development of one. The landscape surrounding the building will contribute heavily to the shelter provided for the building from the wind, the sun and from flooding. It is more than simple amenity and it presents the opportunity to fully capitalise on the investment made in the land the site is on.

Standard practice when carrying out site comparison assessments at early stages often skims over the practice of carrying out detailed ground investigations and ecological value reviews as they are considered too expensive. Nonetheless, investing in the comparatively small costs of these assessments early on is paramount when considering the whole life impact of a project. Costs for investigations that are in enough detail to make informed decisions constitute less than 0.1% of overall project costs and yet a bad decision can affect over 10% of the overall project costs in the long run.

The choices made with respect to the landscape affect the perceived quality of the project. The landscape also provides the route to join up the site with its surroundings and create or contribute to existing biodiversity, ecology and wildlife and sustainable urban drainage strategies. Again, in this context, early decisions provide significant added value with additional capital cost benefits. Added value is in the form of:

- Simple climate adaptation potential;
- Straightforward, powerful and long lasting placemaking;
- Opportunities for easier planning approval when actions positively affect the environment of more than just the site boundaries;
- Easier and cost effective local and central government planning and environmental legislation compliance.

Land available for building development is an ever decreasing and finite resource. There is considerable evidence, dating back several decades and culminating in the most recent worldwide climate emergency declarations, that we need to protect and enhance biodiversity. Thus, land must be used in the most efficient way possible. Significant and long term damage is done whenever land is appropriated for housing, roads, industry and recreation, without full consideration of diverse tangible and intangible services and values those sites provide. Brownfield remediation and regeneration represents the best opportunity to enhance.

According to European Environment Agency (EEA) estimations there are close to 3 million brownfield sites across Europe. Additionally, these sites are often located within urban boundaries (where land is considered more valuable) and are well connected with respect to transport links and access.

These types of sites can be a competitive alternative to the greenfield investments that are often favoured by developers looking for high profit margins alone and who do not consider the full impact of their actions. Fully sustainable regeneration also encourages innovation by the project stakeholders and helps integrate many aspects of development that are often looked at in isolation creating even further opportunities for added value.

Where considering the ecology particular to the land it is important to assess the ecological value of the site. The focus should be on preserving and enhancing the site's value after project completion. This increases the remedial value of the land as well as minimising the threat to the local ecology.

Insulation

Insulation presents the most basic and cost effective way to keep a space thermally comfortable. Optimising the design of the building fabric helps minimise heating and/or and cooling requirements. This has the benefit of capital cost reductions from downsizing the heating and cooling systems. Depending on climate, good building fabric design can eliminate the need for any energy consuming heating, ventilation or cooling systems altogether.

When considering refurbishments or general building upgrades and maintenance tasks the improvement of existing building fabric has significant, almost instantaneous results with respect to comfort and running costs. Overall, the benefits of good building insulation design decisions can be summarised as follows:

- Reduced energy costs;
- Increased internal temperature control;
- Improved productivity;
- Lower capital expenditure;
- Increased building residual value;
- Easy regulatory compliance.

U-values and R-values

Different fabric elements have different heat transfer properties. Glass and metals, for instance, are typically the least able to retain heat. The ability of fabric to transfer heat is a measured factor expressed as its *U*-value. This is a measure of the amount of energy (Watts) that is transferred through a unit area of material when there is a 1°C difference in temperature between each side (inside and out). The lower the *U*-value the better the material is at preventing heat transfer.

Calculating accurate *U*-values for a given composition of material layers is a relatively complex process that has been simplified by dynamic building simulation software.

The thickness of insulation required for any given composition ultimately depends on the R-value (measured in m^2K/W) where:

$$\frac{1}{U} = R = \frac{t}{\lambda}$$

where
t = thickness of insulation (m).
λ = thermal conductivity (W/mK).

The lower the thermal conductivity of the material, the lower the thickness of insulation needed and the more expensive the product, so a sensitivity analysis will have to be undertaken.

For example, if we take the total target *U*-value for a wall to be 0.14, the corresponding *R*-value is 7.14 m^2K/W. To identify the potential insulation thickness and thus understand what the final wall thickness is to be, we can take the combined non-insulation components of the wall make up to have a *U*-value of 0.9 then the *R*-value would be 1.11 m^2K/W. from there we see that the required *R*-value of the insulation would need to be 6.03 m^2K/W.

Where wall thickness is a significant factor for the design team difference insulation materials can be considered to give the required *U*-value (the thinner they are generally the more expensive), so

- 250 mm thick insulation would need $\lambda = 0.041$ W/mK;
- 200 mm thick insulation would need $\lambda = 0.033$ W/mK;
- 150 mm thick insulation would need $\lambda = 0.024$ W/mK;
- 43 mm thick insulation would need $\lambda = 0.007$ W/mK.

Thus, the sensitivity analysis needs to address the difference in cost for the more expensive material against the difference in net area the building ends up with if a less expensive material is needed at higher thickness.

Numerous building regulations suggest minimum *U*-values for different building elements, the balance of an effective design can only really be assessed by using a whole building simulation approach.

Thermal bridging

Thermal bridges, or cold bridges occur when a part of the building fabric's uniform thermal properties are interrupted. This usually occurs when the building fabric is penetrated by materials with a higher thermal conductivity (such as a steel beam). Bridging also can occur where there is a change in the thickness of the fabric, for example, around openings or where walls meet floors and ceilings.

Thermal bridging leads to uncontrolled thermal transfer such as heat or coolth loss. This can also lead to condensation, which in turn causes pattern staining and mould growth around the bridges. This has an effect on internal air quality, and prolonged mould growth is a cause for several respiratory problems and contributes to the sick building syndrome. Additionally, the effect of moisture and water on the building fabric can lead to deterioration of plaster and paintwork and, in some prolonged cases, structural damage.

Current practice does not assess thermal bridging issues in detail beyond the final checks carried out at the end of the design process as needed for regulatory compliance. There is a mistaken perception that checking, simulating and reviewing the design for bridging is an unnecessary add-on even if any additional cost is marginal and, arguably, can be avoided if the design team collaborates effectively with all stakeholders. Conversely, thermal bridging is very easy to avoid early on in the design process and, in fact, if using a layered approach can be eliminated altogether with minimal changes. If left unchecked however, the costs for repairing the damage caused by uncontrolled heat transfer can be significant.

Servicing strategies

Building services can account for between 50% and 75% of the capital and installation costs of a building. Whilst good design of building services is often undervalued, the impact of badly designed services is significant. When considering that services can take up as much as 15% of a building's overall volume, the importance of good collaboration is evident. Many design teams "tolerate" the inputs of building services professionals indicated by the plethora of buildings struggling to keep comfort conditions acceptable for occupants through systems "tucked away" in out of the way and difficult to reach and maintain places.

Not only are building services unavoidable they are essential and, as such, architects, surveyors, constructors, structural engineers, planners, estate managers and all other project stakeholders need to have a basic understanding of how strategies are developed and implemented within the building.

Building services bring life to an otherwise static building; the servicing strategy chosen for a building dictates air and water movement, incorporates electricity and light and delivers the comfortable environmental conditions desired by the occupants. The building servicing strategy is the main contributor to energy consumption within a building. This is not to say that it's just the boilers and pumps and lighting that consume electricity and fuel, it is the interaction of the building fabric and layout with these systems that dictate whether a building is considered effective and efficient or not.

Thus, the fabric first approach to building design is one of the cornerstones to effective building procurement. Regardless of geographical location the building fabric is there to provide a space that retains heat and/or passively cools down when needed. It is one of the most powerful ways to extend a building's life and maximise its potential whilst minimising running and maintenance costs. Beyond this, the strategies used to service the building are what determines how comfort will be maintained and can make or break the perceived effectiveness of the building design.

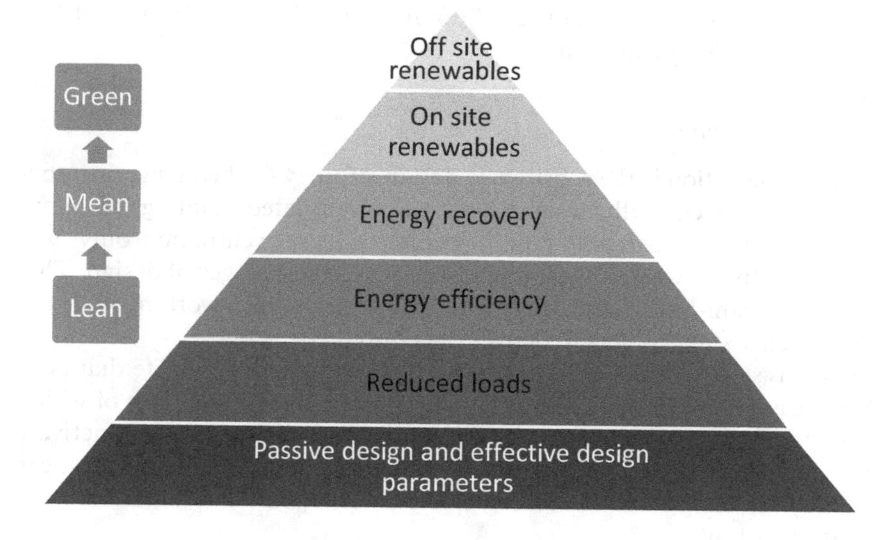

The design and specification of the building fabric has to strike a balance between ventilation, daylight requirements and the need to provide a comfortable environment for the people inside. Building physics and the basics of how humans and the environment interact with a building are essential. Detailed multi-aspect and fully dynamic building system simulation is crucial to determining the best approach for a given building shape on a particular location.

Ventilation

Providing effective ventilation to a building is a primary design requirement; clearly, without it the building is a failure. Over the past 25 years or so building regulations and energy conservation standards have driven designers to continually improve the thermal performance of building fabric and in the same vein designs have sought to dramatically reduce levels of uncontrolled infiltration by increasing building airtightness. This has, indeed, increased carbon and energy savings on the one hand but has increased the risk of overheating (even in colder climates) and poor internal air quality on the other. Now more than ever, there is greater emphasis on correctly designing ventilation systems. Unlike older buildings, newly refurbished buildings and newbuilds cannot rely on uncontrolled infiltration to help maintain internal air quality. The time and effort allocated to assess design, specify, detail, install and commission a ventilation system will dictate its success or failure at building occupation and throughout the building's life. The additional cost of this allocation is generally in the £100s and it is a small price to pay compared to the 10s of thousands of pounds that will be needed to rectify a compromised design later on.

Natural ventilation

Natural ventilation is the most basic design strategy for bringing air into a building. When carefully designed, naturally ventilated buildings are more cost effective to construct and to operate. Natural ventilation only ever works effectively, when it is planned from the earliest stages of design. This requires a team-based approach, with all the disciplines working with the non-designer stakeholders to deliver the desired effect.

There is strong evidence, spanning several decades, to indicate that people have a preference for opening windows and natural light both of which are features of well-designed naturally ventilated buildings. We instinctively reach for "fresh air" in many stressed, charged or emergency circumstances whereby the simple act of moving closer to an open window diffuses the situation instantly.

Although outwardly simple, naturally ventilated buildings are the most complex to design. The process requires considerable effort at design stage with additional time allocated to carry out the necessary studies to combine all aspects of building physics that make a natural ventilation strategy successful in bringing in enough air for good air quality, bringing in enough light for good daylight, avoiding the ingress of distracting external noise, avoiding overheating from high solar gains and removing enough heat to avoid overheating. It's a tall order which is why project teams often are discouraged from pursuing it as an option. The perception is that only small buildings can be naturally ventilated, this is more because small projects attract smaller teams and thus provide more scope for unplanned collaboration rather than anything else.

When designing in silos, mechanical engineers in particular are more often responsible for "erring on the side of caution" and designing in a mechanical ventilation solution (more recently even in non-PassivHaus domestic properties) rather than risk challenging architectural decisions or engaging in collaboration to reach a natural ventilation solution.

Nonetheless, a naturally ventilated building provides the greatest value with respect to secondary and tertiary benefits and the "hurdles" of complexity can be overcome if a conscious decision is taken early on to engage in the fully collaborative approach needed to achieve good results. This is possible for all types of project teams regardless of size, if the brief and requirements are clear.

The benefits of effectively designed natural ventilation can be summarised as follows:

- Lower construction costs – Naturally ventilated buildings can be as much as 15% cheaper to construct than their fully air conditioned equivalents and this saving increases for building over eight storeys high;

- Lower running costs – Naturally ventilated building are up to 18% cheaper to run;
- Lower environmental impact both in terms of embodied carbon and in terms of whole life carbon emissions;
- Increased occupant satisfaction and perception – the building provides an environment linked to the diurnal and seasonal variations in external weather and can more closely conform to occupants' expectations of adaptive thermal comfort;
- Increased productivity;
- Increased building robustness – the process of designing for natural ventilation intrinsically links to a level of in-built flexibility with respect to layout, as well as use type.

There are a number of considerations to be accounted for. For starters, the type, number and location of openings are of paramount importance as is the clear floor to ceiling height of the space. For example, low level openings are good for providing local ventilation for a limited perimeter depth and cannot necessarily be used for deeper floorplans. High level windows are excellent for cross ventilation and trickle ventilation is brilliant for colder weather but aspects such as cold down draughts need to be avoided if comfort is to be maintained.

There are limits as well with respect to building shapes that affect the choice of natural ventilation system design; simple systems can only be used for depths of up to 15 meters beyond which solutions such as atria or passive stacks will need to be employed to help air move around the space by drawing air from or to perimeter areas. For very deep multi-storey buildings however, natural ventilation cannot be the sole solution.

Aspects such as external air quality also require consideration, and if the air outside is too polluted then it is not beneficial to bring it in to the space unfiltered. External noise is becoming a significant issue in cities and the distracting effect of it often outweighs the benefits of natural ventilation; care needs to be taken to assess this against the building's usage patterns and seasonal variations. Internal variations in load also have to be taken into account, such as the occasional and temporary increase in occupant numbers or additional gains, for example, as experienced during special events. Natural ventilation design also has to account for fire regulations and link with fire containment strategies, public health requirements and the specific interactions the building facades will have with the wind (this is slightly more detailed than a simple check against a weather file as wind studies within city streets are far more complex).

The list may seem long and it is an indication of the importance of making the conscious decision to proceed with natural ventilation rather than mistake the simplicity of the solution as an indication that the design process is as trivial a matter.

There is plenty of reference, research and case study material readily available from numerous local and international sources to help with any knowledge and skills gap to achieve a good natural ventilation solution. Design tools have come a very long way in the past 20 years and the accurate prediction of system performance is now consistently reliable if calculations are conducted in enough detail (simplified methods do not work past the very first iterations).

Mechanical ventilation

Where natural ventilation cannot be effectively incorporated into the design, well-designed mechanical ventilation offers the best solution to ensuring good internal air quality by providing enough air to remove internal contaminants and, where necessary, filtration and treatment of external air. The energy consumption of mechanical ventilation systems is often higher than the energy consumption of modern LED lighting systems.

Mechanical ventilation is also the preferred solution where there is a real need for closer temperature control or where the internal heat gains and pollutant concentrations are such that natural ventilation alone cannot handle. Additionally, for instances where building shape precludes the use of natural ventilation or where external noise ingress and security is an issue, mechanical ventilation offers a compromise.

There are three types of mechanical ventilation, extract only, supply only and supply and extract ventilation.

Extract only ventilation is used where contaminated internal air needs to be removed. Examples include kitchen and toilet extracts to remove smells, car park extracts to remove carbon monoxide and leaked vapours, dust and fume extract from industrial applications or even more localised and specialised uses such as hazardous material and fume extraction, etc. Mechanical air extraction provides a constant and predictable ventilation rate and as such is indicated for these uses. Where cross contamination risks are low, heat recovery is also an option on extract only systems.

Supply only ventilation applications are more limited and are generally used to provide filtered and heated air into a space in a more controlled dispersal pattern. It also acts to pressurise an already airtight building so that uncontrolled air infiltration is further reduced.

Supply and extract systems generally use a balanced central air handling unit. Each component of the mechanical ventilation system requires careful consideration. The energy consumption of the system, its effectiveness and, to a degree, its maintainability is governed by the number and type of the components that make it up.

Mechanical ventilation systems consume energy and thus contribute to carbon emissions through the operation of the fan motors. The efficiency of the system depends on the pressure drops in the system and the volume of air that is transported around it. Efficiency is increased when the design reduces pressure drops, uses efficient fans and avoids sending excess air around the system. These three areas are where typical designs fail.

Pressure drops are increased because duct runs are constrained and odd angles across smaller than ideal ducts have to be incorporated. Fan efficiencies are reduced because value engineering exercises fail to identify the cost to benefit ratio of how the saving of a few £10s in capital cost will affect the additional £100s in early replacement and running costs. Air volumes are calculated for extreme maximum occupation and controls are either not installed (see value engineering) or not commissioned appropriately to ensure that air supply is linked to demand.

The main characteristics of these systems and ultimately their main points of failure are:

- System velocity – the lower it is (say a maximum of 3 m/s in ducts) the lower the fan power and its associated noise; however, it attracts higher costs and needs the most space;
- Air flow – good duct design achieves smooth flow with minimal turbulence and very low pressure drops. Such designs mean longer duct runs, the use of in duct turning vanes and avoidance of rectangular ducts, again this requires more space and slightly higher costs but it provides a far better system;
- Filtration – there are different types for different uses and should be selected to achieve the particle extraction effectiveness necessary for the lowest pressure drop, to minimise overall running costs;
- Attenuation – noise in these systems is caused either by the fan or the turbulent air flows through ducts and grilles, if the design has not managed to eliminate the effects of that, then, attenuation will be needed to reduce the noise to acceptable levels both in the system and, sometimes, in the spaces served by the systems;
- Air distribution – air delivery needs to provide a good ventilation effectiveness, avoiding short circuits between supply and exhaust and provide air at a satisfactory speed and temperature in relation to occupant comfort;
- Control – the most effective way of controlling these systems is based on demand, whereby the volume of air supplied to the space is adjusted to meet occupancy. Demand is measured on a combination of:
  - Occupancy through CO_2 sensors or passive infrared (PIR) located appropriately within the space (not in ducts);
  - Temperature and humidity through sensors located in the space or in ducts (depending on type);
  - Overall time switches for switching the systems on and off across the entire zone or building.

Ventilation heat recovery

For buildings that are not to PassivHaus standard heat recovery becomes economic when the carbon and cost value of the recovered heat outweighs the potential increase in capital and running costs. As a general rule of thumb, the viability of heat recovery increases as the number of times per hour that the air volume in a space is changed (referred to as air changes per hour – ach) increases. Its effectiveness increases the higher the temperature difference is between supply and extract. For locations with relatively long heating seasons heat recovery is almost always necessary.

When considered early on, certain aspects of heat recovery can be designed in so as to increase effectiveness. For example, efficiency increases by as much as 40% if supply and extract ducts are adjacent to each other allowing the use of systems such as double accumulators which offer the highest efficiencies with the lowest pressure drops.

Mixed mode ventilation

Mixed mode systems use a combination of natural and mechanical systems for ventilation and regulating thermal comfort. As a solution it combines the best aspects of both systems in an integrated approach, using the building fabric to do as much as possible to create a good internal environment and adding mechanical and electrical systems to do the rest. Skill is needed to design the systems to work with each other, otherwise the result is one where systems clash and create an unpleasant and inefficient environment.

There are three main ways to effectively operate a mixed mode system and these need to be thought out and discussed in advance with stakeholders. The choice of operating procedure in use will also dictate the attractiveness of the system choice when developing a speculative building.

Mixed mode systems can be designed to work together in a concurrent manner whereby a building operates primarily through opening windows but relies on a level of background mechanical ventilation to extract hot or polluted air. Most naturally ventilated buildings with mechanical extract from toilets or kitchens, etc. all fall under this category. Alternatively, the systems may operate in different ways depending on the time of year, occupancy levels, or during different times of the day. This type of changeover operation can be seen in buildings that are naturally ventilated during the day and that utilise mechanically assisted purge ventilation during the night to precool the structure for the following day. Finally mixed mode systems can operate in alternate ways across different occupancies or extreme weather conditions, for example, where a building operates in natural ventilation mode except for when temperatures become too high during unseasonably inclement weather whereby the building is sealed with all windows closed and air conditioning is used.

There are many advantages to the mixed mode approach:

- It provides the occupants with a much greater level of control increasing their satisfaction with and enjoyment of the building;
- It is more energy and carbon efficient than an all mechanical system;

- It is more predictable than a fully naturally ventilated building;
- It creates a more robust building that can still operate acceptably using a much wider variety of operation modes to adapt to changing conditions both in climate and in occupancy;
- It increases flexibility and the building's enduring value and life span;
- It stretches the plan depth limits set by a natural ventilation only approach;
- Maintenance complexity and costs are generally lower;
- It is a more cost effective approach, with mixed mode buildings costing up to 10% less than their fully air conditioned equivalents.

Mixed mode systems require a fully collaborative approach in terms of the design team; designing in silos often leads to unnecessary complexity and reduced effectiveness. In addition, all design and operational intents need to be made clear, using simple language and easy to understand schematics, to the building management, to individual occupants, to operation and maintenance contractors and to those who may adapt, equip and alter the building, including space planners and interior designers.

Effective commissioning is paramount so that systems are controlled, checked and tuned in line with the design intent. To avoid the pitfalls of aspects falling through the cracks, the systems themselves need to be robust, adaptable and made as easy as possible to set up and to alter as necessary and as the building use evolves over its life span.

Lighting

Over the years lighting design has gone through a number of "fashions". For a period of about 30 years the main desire was to create an internal environment that was stable and unchanging regardless of external variations in conditions. The main driver for this was business and commerce in general with little regard for the effect this may have on human wellbeing. There is widespread evidence of these practices in the thousands of buildings across the world that resemble and "feel" like battery farms, with building spans reaching over 30 meters with little access to natural light, or fresh air. More recently with our increasing understanding of the effect of the internal environment on our health, the focus has been on the provision of natural light and views. The past 30 or so years have seen the proliferation of buildings with floor to ceiling windows creating huge class boxes increasing the need for cooling even in colder climates. In both cases, the emphasis has been on large occupancy densities for commercial and office buildings and the decrease in manual lighting controls.

Good lighting design needs to strike a balance between these extremes, taking into account the limitations of automatic electric lighting controls, the capabilities of occupants to override automations to achieve adequate levels of comfort and lessons learnt from our past successes and mistakes.

There is considerable material available to help project teams reach an effective design and the skills needed to deliver a good solution should be valued by stakeholders. The following sections provide information on the overall benefits of good lighting design with some indications of the consequences of mistakes.

There is a plethora of material to be used as guidance for good lighting design, and this section touches on some of the main aspects of this with respect to the level of detail required. Research papers, good practice guides and basic legislative requirements for internal and external lighting design are continually updated to account for the rate of technological and research advancements and are available from professional organisations such as the ISO and should be referred to.

Natural lighting

Daylighting should be the primary source of light energy for the building. When considering daylight in simple energy terms the benefits are generally constrained to energy; carbon and cost savings with respect to reduced electrical energy consumption for lighting, cooling and heating loads are also affected; however, the impacts reduced cooling loads in summer and increased heating loads in winter tend to cancel each other out. With the technological advances that have meant significant improvements in electric lighting efficiency, the electrical savings in the use of daylighting have diminished by close to 90% in the past 20 years. As such, these savings alone are indeed minimal and considered insufficient to make a case for improved daylight design.

Where justification for the widespread use of daylight is needed the benefits need to be considered with respect to occupants. Daylight availability significantly impacts occupant wellbeing and productivity. When done well the results are considerable, when done badly the impact is substantial, in both cases there are financial implications to the organisations occupying the space (workplaces, schools, healthcare facilities, etc.) and in domestic situations occupants feel a sense of wellbeing that translates to measurable results in terms of improved sleep and overall psychological wellness.

Daylight is one of the most effective stimulants to the human visual system and to our circadian rhythm. Disruptions to our circadian rhythms cause verifiable health problems beyond just a reduction in the effective performance of tasks at work or at home. Natural light also affects mood for the relatively short term.

Observations show that moving occupants from a space with low daylight levels into a more naturally lit space will improve mood until individual expectations are adjusted. Conversely, the opposite occurs when moving occupants from a well daylit space to a less daylit space hereby their mood will deteriorate until individuals get used to the reduced levels. This trait is most effectively capitalised upon in the retail environment whereby it has been proven that well daylit retail spaces have higher sales volumes than the same spaces with no daylight.

Good daylighting can maximise visual performance better than most forms of electric light because it is generally delivered in large quantities and provides excellent colour rendering.

Views to the outside world also provide benefits with respect to measured reduction in stress for occupants and this has been linked to our innate human need to be in some form of contact with nature and the seasons.

TECHNIQUES

Inexpertly designed buildings often have glare and rather than providing benefits, daylight provides uncomfortable distractions through veiling reflections and shadows, although electric lights can have the same effects designers are more equipped to design electric lights better than designing for useful daylight. Occupants will often take quick action to eliminate daylight if it causes consistent discomfort or increases task difficulty. The most common example of this is the practice of "blinds down lights on" that can be experienced in hundreds of buildings across the world giving stark evidence of the level of underutilisation of daylight across the buildings trade.

There are some techniques that skilled designers can use to avoid these problems effectively. Primary among these is the use of Useful Daylight Index (UDI) or Useful Daylight Illuminance (also UDI) or climate based daylight assessments as a complete replacement to the prevalent Daylight Factor (DF). The UDI analysis acknowledges that daylight illuminances in buildings vary enormously throughout the year and are closely linked to both climate and the individual building's surroundings.

The daylight factor is the most commonly (99%) used metric not because of its accuracy but because of its simplicity. It is the ratio of available light under bright overcast sky conditions that enters a space and, in many cases, it is the only factor considered when designing for daylight. As such designs tend to be problematic, because they are based on false and simplistic assumptions. However, for years DF has been used as a metric by almost all general guides and overarching briefs in certification schemes such as BREEAM and LEED as well as in guides by organisations such as the British Council of Offices (BCO). Within the past 2–3 years, even these guidance notes and certification schemes are cordoning on to the fact that UDI is a far more effective measure of daylight availability and are slowly incorporating climate based assessments into their requirements for compliance.

The use of detailed performance evaluation of daylight at the early design stage is not only recommended but should be common place. Software and technological advances as well as detailed datasets are vastly available to accurately and reliably predict the behaviour of daylight in a space based on time varying climatic data for any given location, orientation and azimuth. Adding information to such models like the position and importantly the external finish or surrounding building and obstructions further enhances the accuracy of such studies. This is especially useful if a building is surrounded by tall fully glazed reflective surfaces, as is most common in city centres across the globe.

Simulating for accurate internal daylight distribution and illuminances representing realistic sun and sky conditions for a variety of façade and opening options is not only valuable but can mean significant savings as, 9 times out of 10, it means the need for relatively smaller windows compared to daylight factor assessments.

Ultimately good design needs to provide adequate useful light for the given use of each space (offices will have different requirements than retail spaces or domestic properties) for at least 50% of the operating hours of a space.

Instances of occupant discomfort are predictable with the right analytical approach. High levels of illuminance, or too much light (say 1,000 lux) need to be limited to a small percentage of the space and to an agreed maximum; for example, 10% of the space for less than 250 hours per year. Increased heat gains leading to overheating also need to be limited to an agreed maximum; for example, a limit on heat gains from solar beam radiation (insolation) can be set to 25 W/m^2 at any time during occupied hours outside of the heating season.

Well designed buildings, with accurately assessed daylight performance, generally, will not need internal blinds and the instance of blinds-down/lights-on will, simply, not occur.

Artificial lighting

Although technically, daylight is simply a source of electromagnetic radiation and electric lights can be designed and specified to match specific light spectrums to mimic daylight, they cannot fully match the variations in daylight spectrum that occur over the course of a day, or throughout the seasons or in response to different weather conditions. As such, the effect of daylight cannot be fully mimicked; nonetheless, there are some basic conditions that define good artificial light provision for when daylighting is not available.

Ultimately, energy efficiency is an easy measure of assessing design performance and simple calculations early on give an indication of just how effective the design is. Aspects such as overall lamp efficiency can be assessed in lumens (how much light is emitted) per circuit watt (how much energy is needed to each lighting system). A good lower limit is 80 lumens/Circuit W across all fittings in a space (to account for non-uniform installations), and going below this is an indication that sub-standard fittings and/or lamps are specified which will waste energy unnecessarily.

Another metric is load density measured in watts per meter squared (W/m^2), 4 W/m^2 is easily achievable with today's lighting technology, only 20 years ago this was as high as 20 W/m^2.

If a dynamic climate based assessment for natural lighting is undertaken then aspects such as surface reflectance will already be known or logical assumption will be made, these values are also helpful when assessing artificial lighting designs as they provide a measure of how the light bounces off ceilings, walls, floors and furniture and affect aspects such as brightness contrasts.

LIGHTING CONTROLS

Lighting controls, and their complexity, have often been a bane for designers and occupants alike and in many cases rightly so. When it comes to controlling lights simplicity is key. The value of simple, well placed and well labelled switches is immeasurable but can easily be quantified compared to all singing all dancing automated systems that on the whole rarely work to plan. Control really needs to account for a couple of situations:

- When the space is not occupied, lights should be off;
- When the space, or part of it, has enough natural light, lights should be off.

By simplifying this part of the design, time, effort and some of the released capital can be spent ensuring that the behaviour of natural light entering the space is accurately assessed.

External lighting

Outdoor lighting should be provided primarily to allow occupants and users to perform outdoor activities efficiently and accurately. This includes the use of light to increase the sense of security people need to feel at night in order to consider a space safe and attractive. Light is also a very real way of creating a sense of place within a community and is often used to increase footfall to buildings.

Different light levels are needed for different tasks so light needed for simple wayfinding or signage, is very different than lighting for security or sporting or leisure activities.

External lights require relatively different design considerations than internal lights do as they operate for longer hours and are more at risk from the elements and require more complex maintenance. Here energy savings, eliminating light pollution and ensuring easy replacement are the driving factors.

Controls can still be kept very simple. Simple time switches linked to daylight sensors are the most common. Effective presence detection is important for security purposes.

Heating

In most buildings, heating and hot water account for a significant percentage of energy consumption and subsequent carbon. There is an extensive range of heating system design solutions available to meet the operational requirements of a building, and depending on the type of building there may be a mix of systems. Choosing the optimal energy efficient and cost effective solution requires careful consideration.

It is highly recommended that full dynamic building physics simulation is used to decide on the optimal solution for a building through a thorough options appraisal of alternative techniques.

Heating systems can generally be split into three main subsystems. The primary heating system includes the fuel and heat source; the secondary system includes the heat distribution components and the control strategy whilst the tertiary system comprises the heat emitters. Each sub-system has its own intricacies.

Primary heating system (heat generation)

FUEL

One of the initial decisions to be made is on the choice of fuel. In many cases this will be determined by practical reasons, such as the availability of certain fuels on site, space availability, storage costs, etc. Whatever the choice, all fuel types have an associated carbon impact which reflects the losses from point of extraction or generation to delivery for use on site. This is why, for example, in the UK, natural gas currently has a smaller carbon impact per metered kWh than electricity.

The carbon impact of the heating fuel will also be affected by the efficiency of the heating system chosen. The CO_2 emissions attributable to each fuel type are typically calculated annually and generally are expressed in units of kg of carbon per kWh of primary energy. These figures can be discussed at early stages in the design with stakeholders to help make informed decisions. Currently in the UK natural gas still produces the lowest CO_2 emission level per unit of heat delivered.

The most common fuel sources are:

- Natural gas – methane from the gas mains. It is the most prolific and carbon efficient fossil fuel available in the UK and it is relatively efficiently converted into heat.
- LPG – liquid petroleum gas, normally propane or butane used generally where natural gas is not available whereby it will give similar efficiencies to those systems operated on natural gas. However, LPG is more expensive than natural gas and the additional cost of storage tanks (which are usually rented from the fuel supplier) needs to be considered.
- Oil type C2 – kerosene or oil type D – “gas oil” oil-fired boilers are common alternatives in older constructions and, where mains gas and LPG are not available and are capable of giving comparable running costs, the additional cost of storage tanks and access for re-filling needs to be considered. Carbon emissions are about twice that of mains gas.

- Solid mineral fuel – coal or coke. This option is of exceptionally low efficiency and produces significant local air pollution and it is generally used for much smaller installations and is not a preferable solution as a main source of heat for the building. Large storage and access is necessary, and there are additional requirements to prevent dust pollution during re-fuelling. Carbon emissions are about three times that of mains gas.
- Biomass fuel – wood logs, pellets or chips. Pellets are the most efficient of the biomass fuels for heating, and the carbon impact depends on the pellet source and its travel distance from source to the building. Biomass supplied from around a 30-mile radius would constitute a super low carbon solution. Again, significant storage is necessary as is access to the site for frequent deliveries and additional air filtration will be needed for the heating system to avoid local air pollution from combustion of biomass fuels.
- Electric – electric heating sources are the most efficient as 100% of the electricity can be converted to heat; however, the carbon impact of using electricity is still relatively high being about four times that of mains gas for most countries because of the inefficient ways in which it is produced. In countries where electricity production is primarily by renewable sources (over 70%) and thus where the carbon impact is lower than that of natural gas then this can be a very good fuel source.

HEATING PLANT

Commonly, the heat source for a building will be a boiler providing hot water at specific temperatures to be distributed and used in different ways to heat spaces and hot water for consumption. The efficiency of the heating plant is the major factor affecting the energy efficiency of central heating systems.

Generally, boilers are provided with a quoted efficiency which refers to full or maximum load conditions. This is when they operate at their best and is very different from their average installed efficiency which is more representative of how they operate across a year. Heating plant operates, for the majority of the time, at less than full load because these maximum load conditions are rarely met in real life across the span of the year. Designers should take into account boiler efficiency across the complete range of likely loads as efficiencies can considerably reduce (by as much as 40%) when the system operates at part load. Consideration should be given, and information should be required from manufacturers on values that represent the seasonal average operational efficiency.

NON-CONDENSING BOILERS

Boiler plant is sized (often too generously to allow for contingencies and design errors) for extreme winter requirements, giving it considerable overcapacity for the rest of the year. Efficiency drops rapidly as the load falls and can average to something like 65% over a heating season for traditional designs, this can be even less where sizing is not as tight and controls not as effective.

Most modern non-condensing boilers (through legislative changes) can achieve higher efficiencies by minimising casing and sensible flue losses. That only covers part of the problem, and the time spent running at part load needs to be reduced as much as possible. One technique is to use multiple smaller boiler units linked together as modular boilers instead of a single large one.

CONDENSING BOILERS

Condensing boilers perform better than non-condensing boilers because they are designed with a secondary heat exchanger used to extract additional (otherwise lost) heat, from the exhaust gases in the flue. With natural gas they have seasonal efficiencies of between 83% and 92%. Typically, a condensing gas boiler would have a 15% better seasonal efficiency compared to a new non-condensing boiler or 30% compared to an older type boiler.

Within the limits of each boiler, the lower the return water temperature the more latent heat can be extracted, giving better performance. This can lead to higher efficiencies at part load than at full load, hence the preference now for condensing boilers. However, this means that the designers will need to take this into account and design specifically for a condensing boiler operation, which is different than traditional design requirements.

There are solutions whereby a multi-boiler installation includes condensing and non-condensing boilers. In these cases, the condensing boiler should be the lead boiler. This will ensure that the boiler with the highest efficiency runs for the longest number of hours and that full advantage is taken of their condensing capability during part load. Condensing boilers are ideal for use with weather compensated heating circuits and underfloor heating. Other potential uses include domestic hot water, where there is a high load, or swimming pools.

COMBINED HEAT AND POWER

Combined heat and power (CHP) systems provide the opportunity to produce both heat and electricity on site. Overall a CHP system produces heat in a less efficient way than a condensing boiler; however, it offers environmental benefits through reduction in emissions of the principal greenhouse gas CO_2 for electricity generation, and also the emission of SO_2, which is a major contributor to acid rain; thus, the balance of heat and power produced means that they are considered as more carbon efficient when considering both heat and electricity. They are at their most cost and carbon effective when there is a constant year round need to use all of the heat and electricity produced.

Oversized CHP systems tend to be extremely inefficient and tend to become very expensive to run. When considering a CHP plant, it is vital that all possible steps have been taken to make the building as efficient as possible and thus the heating load as low as possible. This will help to avoid installing incorrectly sized plant. Future changes in energy requirements should also be considered, especially the possibility of reductions in heat or power demands. As a rule of thumb, applications which have a simultaneous demand for heat and power for more than 4,000 hours per year will be worth investigating in detail. Depending on the requirements, supplementary electrical power will be required from the grid at times of peak use.

Applications of CHP for building services generally use small-scale CHP. These units have electrical outputs of up to about 1 MWe and usually come as packaged units based on gas-fired reciprocating engines, with all components assembled ready for connection to a building's central heating and electrical distribution systems. Alternatively, micro CHP systems which are slightly less efficient are available that provide enough heat for domestic-type heating and hot water systems and are rated at about 1 kWe.

SOLAR HEATING SYSTEMS

Solar water heating systems are well established as a reliable and accessible way to actively use renewable solar energy. A common size of solar panel for domestic use is about 3–4 m^2 will deliver about 1,000 kWh per year in the UK. As most of this occurs in the summer months, it is necessary to supplement solar heating with a conventional boiler or other system. A typical solar water heating system comprises collectors and a water storage cylinder with connecting pipework. Some systems use thermosyphons (the natural buoyancy of the heated water) to circulate the heating water, but most UK systems need a pump because of low temperature difference.

If not heating water solar heating systems can be used to preheat parts of the overall system such as ventilation air. An example is where cold air is drawn from outside and passed through a specially designed part of the building such as south-facing double skin whereby the air is warmed up before it enters an AHU to be conventionally heated (if necessary) before being distributed to the space.

HEAT PUMPS

Heat pumps use electricity to take heat from a relatively low temperature source and discharge it at a higher temperature for use in the building. A heat pump is, in effect, a reversed refrigeration system. There are four essential components and a circulating refrigerant provides the heat transfer. A compressor pressurises the refrigerant gas and pumps it to a condenser where it cools and releases its heat to either a water circuit or to air. The condensed refrigerant, still at high pressure, then passes through an expansion valve where the rapid expansion induces cooling. The drop in temperature liquefies the refrigerant, which passes on to the evaporator. Here the cold liquid evaporates as it picks up heat from the surroundings. A typical evaporating temperature is –2°C, so it can readily collect heat, even in colder times of the year. Unlike other heating systems heat pumps can also be used for cooling.

Heat pumps are most efficient when they produce hot water at lower temperatures, typically around 40°C; therefore their applications can be limited if not designed effectively. The efficiency of a heat pump is measured by its coefficient of performance (COP). A typical value is about 3.0. This means that for every 1 kWh of electricity used to drive the compressor, 2 kWh of "free" heat is collected from the evaporator and 3 kWh of useful heat is delivered for use. The source of heat can be air, water (like rivers, lakes, the sea or even waste water if it is in large amounts) or the ground. Higher source temperatures lead to higher COPs, so air source heat pumps (ASHPs) are the least efficient with typical seasonal COPs of 2.0–2.5, whereas well controlled ground source heat pumps (GSHPs) are more efficient, with potential COPs of over 3.0.

ASHPs are the easiest to install. The efficiency of these systems is inherently linked to ambient air temperatures. In winter when providing heat the system is less efficient when ambient air temperatures are lower. In summer when providing cooling the system is less efficient when ambient air temperatures are higher. Air source heat pumps therefore operate best in environments with long mild mid seasons and do suffer from icing of the evaporator coil in winter. The energy required to melt this further reduces the COP. Air distribution fans on both evaporator and condenser coils add to the system energy overhead. In general, as an alternative to gas careful consideration is needed to ensure that the COP is high enough throughout the year to provide effective overall carbon savings. They do, however, offer a good alternative to other fossil fuels such as oil or solid fuels as they require much less space and can provide comparable carbon savings.

Water source heat pumps (WSHPs) are as effective as GSHPs but require suitable locations with enough space for the coils and an adequate water source where their operation will not affect existing wildlife and, as such, are much less common.

Ground sourced heat pumps extract heat from the ground and pump it into a building to provide space heating and to pre-heat domestic hot water. In the warmer parts of the year this process can be reversed whereby the heat pump rejects heat to the ground, to meet the cooling requirements of a building and to "re-charge" the ground for the heating season. The technology relies on ground temperatures remaining at a stable level of between 8°C and 14°C. Where systems are not well designed the ground source coils can suffer from icing significantly reducing their COPs and depleting the stores of energy in the ground which can create long term problems for the system and the building.

DISTRICT HEATING

Community or district heating (DH) systems generate heat from one or more energy sources and deliver it to buildings and users via distribution pipes. DH schemes can normally be found in domestic or mixed use developments and are very prevalent in northern Europe and Russia. Systems that fall under this category can vary in size from simple, apartment buildings fed with from a communal boiler, through to large sites where several hundred domestic units and/or education and commercial buildings are supplied with heat from a central energy source, all the way to city-wide heat distribution networks that serve a wide range of customers.

Projects generally either connect to an existing system or develop a district heating system as part of the works. Commonly, a central energy centre is required that produces heat and or electricity for local distribution. Distribution is normally via pre-insulated pipes to minimise distribution losses. The type of pipe (steel or plastic, flexible or rigid) is dictated by the system type and size. Heat transfer between the distribution network and the building or dwelling's internal heating distribution systems occurs via a hydraulic interface unit (HIU). These normally include a heat exchanger(s) which serves as the substitute for the individual boiler. The DH network usually is also used to provide DHW; therefore removing the need for hot water cylinders or any type of individual boiler at a dwelling or building level also frees up extra space.

From the end users' point of view, there is no operational difference between this type of system and an individual heating installation; the secondary and tertiary systems are the same regardless of the heating source.

District heating systems and networks are straightforward concepts that require considerable care and expertise in their design, specification and installation. There are many points of failure in these systems and the main considerations are the dynamic heat loads expected from the system over the course of at least one year and the heat density of the network that they serve. Additionally, the management, operation and maintenance of the network present complexities that have to be ironed out at the very beginning of the project. There is considerable data available on design and operation of DH systems and on the design of the systems that tap into them. In all cases a specialist should be employed and detailed assessments taking into account accurate energy consumption and generation predictions should be completed.

Ultimately, the cost-effectiveness of new DH systems depends on the ability of the centralised heat production to outweigh the disadvantages relating to heat distribution. The carbon effectiveness of DH systems depends on the type of fuel or combination of fuels used to generate and potentially store heat centrally and the associated distribution losses. DH systems are, however, very well suited to provide for multiple renewables led fuel types allowing for significantly higher security of supply than individual single fuel systems and, thus, can be considered viable solutions to the environmental crisis the planet is currently facing.

HOT WATER PLANT

A well-designed hot water system is characterised by lower energy and water use; lower maintenance costs and improved water hygiene. Hot water can either be generated centrally (and pumped around the building), or locally at the point of use.

Domestic hot water (DHW) is required at lower temperatures (typically 60°C) than water for heating (typically 80°C).

Where a central boiler plant provides both heating and DHW, summertime operation for DHW can be very inefficient because boilers that are sized to deal with the winter heating load and so have low efficiencies when operating at the low loads needed to provide DHW only. Also, at lower outputs the distribution losses increase as a proportion of the total heat demand.

To improve efficiency and increased system resilience, the installation of dedicated DHW plant should be considered. The plant will operate efficiently throughout the year as DHW water loads tend to be steady for a given building type with minimal distribution pipework and associated losses. Where the requirement for hot water is small and isolated, it may be worth installing instantaneous point-of-use water heaters, thus increasing efficiency by avoiding heat loss from pipework and storage cylinders.

There are some additional considerations with respect to DHW. A typical design will allow for a level of hot water storage (to smooth out intermittent use issues that decrease plant efficiency and lifespan). Thus, it is important that the size of storage matched to demand. Excessive storage volumes lead to high standing heat losses and also increased risk of bacteriological infection of the water stored. Where the plant room areas are limited or depending on heat source plate heat exchangers (PHE) or copper fin water heaters may be an option. These systems transfer heat more effectively than traditional storage-type water heaters and although they are more expensive, they are more cost-effective where larger quantities of hot water are required, for example, for process use, and reduce the requirement for storage.

Water softness needs to be considered as well, and in hard water areas scaling may occur which will affect the efficiency, the life span and maintenance requirements for the plant; in these cases water treatment in the form of softeners may be needed. This may increase capital costs marginally but will pay back considerably over the life of the system.

Secondary heating system (heat distribution)

PIPES AND DUCTS

Minimising distribution losses maximises system efficiency and improves control effectiveness; it also keeps consumption low.

Heat distribution systems are generally either pipework or ductwork. The main point is to ensure that the runs are as small as possible to avoid excess losses by design. And, in the case of DHW, long distribution runs also create dead legs where water can stagnate; thus careful design is recommended.

Once runs are minimised all distribution pipework and ductwork should be well insulated. As the building itself will be well insulated any uncontrolled heat gains can become problematic especially where pipe or duct run are located, unwanted heat in those areas can cause secondary problems such as increasing the temperature of cold water supply pipes increasing the risk of legionella posing a real threat to occupant health. Effective, fully installed, gap free pipe and duct insulation helps alleviate this to a great degree.

To enhance effectiveness and prolong life adhesive heat-resistant foil tape should be wrapped around the insulation. It protects the main layer from damage and further helps to prevent heat loss. This tape is often disregarded and left uninstalled. The potential savings from its omission generally come to pennies compared to overall costs and but the benefits in terms of insulation effectiveness and the increase in longevity far outweigh these savings. Generally, contractors that omit to use the tape simply provide indication of bad workmanship rather than cost effective construction.

HOT WATER STORAGE

Heat will be lost from storage vessels even when they are insulated; these are called standing losses, and good designers are able to predict what these losses are going to be and minimise them. As a general rule of thumb, the greater the volume, the higher the heat loss. Therefore, hot water storage, regardless of its use, should be kept to the minimum acceptable for the demand.

When considering domestic hot water, it is essential that requirements for the control of Legionella are not compromised. All relevant legislation and recommendations should be taken into account.

FANS AND PUMPS

Fans and pumps are used to push warm air or hot water around the distribution system. Their specification, detailing, installation and commissioning are all vital parts to ensure effective system operation.

For systems to be energy efficient the systems need to be designed so as to reduce as much as possible the load on these pumps and fans, after which, the specification of the most efficient motors is key. As a general rule, direct drive fans and pumps are more efficient than belt driven alternatives and can be more straightforward to maintain.

Depending on the operational requirements, the size and type of motor powering the fans or pumps, variable speed drives may provide additional benefits and better system control by varying the speed of the motor. For water pumps successful integration of variable speed drives lies in the valving, systems sensing and control techniques. Where these are not well defined the system is not cost effective. In all cases, care should be taken to ensure that the pump motor and its associated drive are specified together and are matched appropriately.

With respect to fans variable speed drives specification will also dictate their component specifications across the system. As such, it should be considered holistically as it applies across the air distribution system. Again, here, inverter specification should account for fan type, the type of in duct dampers and distribution grilles and importantly the type and location of sensors needed to effectively control the fans.

Tertiary heating system (heat emitters)

Radiators and convectors are the principal means of heat emission in most buildings. Less popular alternatives include exposed pipes and radiant panels for use in warehousing, workshops and factories, where aesthetics is not an issue. The following is a non-exhaustive list of the most common types of heat emitters found.

RADIATORS SERVED BY BOILERS

This is the most common method of delivering heat to a space and it uses a boiler to heat water and deliver it to radiators located in appropriate places within the space. Despite the name, less than 40% of the heat is transferred to the space by radiation. The rest of the heat is actually delivered by convection.

CONVECTORS SERVED BY BOILERS

Convectors sometimes look like bulky radiators where they are not incorporated into the internal finishes within a space and they have a steel casing containing a finned heat exchanger. About 90% of the heat is transferred by convection and this is sometimes enhanced with a thermostatically controlled fan located within the casing. They are more effective than radiators for heating larger areas.

AIR HEATING

Just like the radiator system, a boiler delivers hot water to a heater battery within an air handling unit. Warm air is supplied to the space via air diffusers or grilles with adjustable louvres; generally, the finish is flush with the ceiling or floor.

DIRECT ELECTRIC HEATING

This can either be in the form of electric wall heaters, night storage heaters or warm air heaters, and all are located directly in the space they are heating and do not require a central plant.

UNDERFLOOR HEATING

Underfloor heating generally consists of pipework panels embedded in the floor screed. Heat distribution is uniform and the system provides a high standard of thermal comfort as heat is emitted from the building fabric. However, these systems have a slow thermal response as fabric takes time to heat up and to lose its heat as such they have to be controlled differently and thus separately from other quicker response systems such as radiators.

Additionally, the temperature of water required for these systems from the boiler is up to 30°C lower than that for radiators, as such special consideration is needed in how these systems are incorporated into the overall servicing strategy for the building.

Cooling

Comfort cooling is required in buildings whereby building design quality and climate lead to unacceptably frequent occurrences of overheating. Before incorporating refrigerant cooling, project stakeholders should investigate all feasible options to incorporate "free cooling" into the building. This helps to reduce the investment required for cooling equipment and provides further building robustness and flexibility.

All design considerations need to be assessed using dynamic building energy simulation. The overall design, layout and operation of the building affect the need for cooling. The level if heat gains from internal and external sources also contribute and in many instances, external shading devices can eliminate the need for cooling if designed effectively taking into account the suns trajectory and all other weather conditions over the period of a year. Where these options have been exhausted, then comfort cooling using refrigerants and air conditioning can be considered.

Cooling systems are, of course, specialised design items and need to be considered collaboratively by the project team and designed and detailed by building services engineers. There are many cooling solutions available depending on building design and some of the more common ones are listed below:

- Night cooling with or without mechanical ventilation support;
- Slab cooling using air or water;
- Ground or water source heat pump cooling;
- Evaporative cooling with or without desiccant dehumidification;
- Absorption cooling;
- Chilled beams;
- Displacement ventilation;
- Conventional refrigerant cooling.

The following is a general description of the main characteristics of conventional refrigerant cooling systems.

Refrigerant cooling plant will most likely use vapour compression chillers which utilise HCFCs or HFCs as refrigerants. These refrigerants are less environmentally harmful than CFCs in terms of their ozone-depleting potential and they still have an appreciable global warming effect if released into the atmosphere and are a thousand times more harmful than CO_2.

If these refrigerants cannot be avoided, appropriate safeguards should therefore be taken to prevent leakage throughout the life of the plant.

Designers should be encouraged to design for the highest efficiencies possible and this will dictate to a great degree the location of the chillers, relative to the AHU and relative to the building.

Centralised systems

These air systems treat air centrally within an air handling unit served by a chiller. Chillers (much like boilers) deliver chilled water around a network of pipes to cooling coils in the air handling unit. Depending on the system chillers can be air cooled or water cooled (cooling towers). The treated air is then distributed around the building through the ventilation ducts.

Common centralised systems are broken down into constant volume (CAV) systems, which become complex when used with zones. Variable are volume (VAV) systems are similar to CAV systems but allow for easier demand control across the building. They are more expensive but provide far greater system flexibility and energy efficiency.

The cooling effectiveness of displacement ventilation systems can be enhanced by chillers, and in many cases provide a good middle ground between a passive cooling technique and the consistency of using a refrigerant cooling system. Chillers needed for displacement ventilation systems are generally smaller than those needed for CAV or VAV systems because they inherently make as much use of passive natural convection and free cooling as possible before needing input from the chiller.

Partially centralised systems

These air systems partially treat air centrally within an air handling unit; however, they provide a second level of treatment at local level. If a CAV system is used this provides slightly higher local flexibility. The secondary treatment is typically through fan coil units (FCUs) which in essence act as micro air handling units generally located on the ceiling. These can be controlled individually or in groups and so lend themselves well to zoned control systems, however, overall system efficiency decreases as air volumes increase.

Chilled ceilings or chilled beams fall under the category of partially centralised systems as they need the chiller to provide chilled water but do not require an air handling unit as they act as reverse radiators and do not circulate air.

Unitary systems

Unitary systems are not linked to any central plant and provide cooling in their immediate vicinity. Most unitary systems do not provide ventilation rather treat the air by recirculating it within a space. Split systems fall under this category whereby an internal unit treats the air and a closely linked external unit houses the refrigeration equipment.

Variable refrigerant flow (VRF) systems are slightly more effective as they allow for one single outdoor unit to serve multiple indoor units by varying the refrigerant flow to each internal unit. Additionally, if VRF systems are used for heating and cooling they can be considered as air source heat pumps and can offer some energy savings.

Unitary systems generally require additional design to provide ventilation in to the space, and as control is not easily centralised, require careful consideration as the risk of excessive energy consumption is greater.

Controls

Controls are needed to ensure that the heating systems only operate as and when necessary. The main components for controlling a heating system are the system zoning strategy, time, temperature and boiler controls. The operation efficiency and effectiveness are both governed by how the system is controlled and maintained. When designed, detailed, installed, commissioned and maintained correctly controls can increase system lifespan, improve effectiveness and contribute greatly to reducing running costs. In many instances the difference between a building that costs the earth to run and a super low cost building is in its controls. Occupant satisfaction is also directly linked to how well a building is controlled, and the perception of a good building is linked to how comfortable individuals feel within it.

Good controls tend to seem expensive because the impacts are not fully understood. But the investment still needs to be made in terms of time during design, collaboration across the design team and other stakeholders as well as time during commissioning and handover. The cost implications of control installations are marginal compared to the benefits with respect to plant and building longevity, and the greater the time investment at the start the lower the additional capital costs can be.

The consequences of poor control design can be significant with additional, often unplanned, time needed at handover and commissioning. Badly controlled systems develop problems and require emergency breakdown maintenance often. Occupants become unhappy and the building does not perform as the developed and designers intended with respect to energy or comfort. In some cases, health issues such as legionella infections arise and this ultimately can lead to litigation. The additional costs of these consequences can be in the tens of £1,000s. For building owner occupiers this should be significant enough to ensure that adequate time is allowed at design for suitably experienced professionals to carry out effective studies. For developers the reputational consequences of effectively selling lemons as peaches can also be significant.

Design engineers are most qualified to understand the interactions between control equipment and operating strategies; as such this aspect should be given as much attention during stakeholder engagements and meetings as would the aesthetics or layout of the building.

The following provides some limited information into the main types of controls available and how, in general terms, they operate.

Time controls

Time controls offer the simplest control strategy available. In essence, they are there to switch systems or individual zones on or off at particular times. Seven day controllers are the most effective providing flexibility with respect to changes of use for the building or space for periods such as weekends or for planned special events. Seven day controllers are also available for electric heating systems and should be used to ensure maximum efficiency.

More advanced time controllers also offer the opportunity to allow for extended warm up or cool down times if necessary. Optimum start controllers offer this flexibility and link time controllers to temperature and they are applicable to all types of heating systems. This type of controller monitors internal and external temperatures to ensure that the heating is on for the minimum amount of time to achieve the desired comfort temperature at the start of occupancy. Optimum stop controllers switch heating off as needed to maintain comfort conditions until the end of the occupancy period and this often means that systems are turned off before everyone leaves the building and the system is allowed to gently cool down at the end of the day. In well insulated buildings this can shave hours off the operation of the boiler.

Temperature controls

ROOM THERMOSTATS

Room thermostats, typically placed on a wall, feed back to a central control panel. Depending on the heat emitter different thermostats are needed and as such should be specified by a specialist. Care is needed for detailing and installation as all thermostats need to be clear of obstructions and be situated where their reading is not affected by draughts, direct sunlight, heat from radiators, office equipment, etc. so that they do not provide false readings.

THERMOSTATIC RADIATOR VALVES

Individual radiators may be controlled via thermostatic radiator valves (TRVs) directly placed on radiators. Again, placement must be appropriate to avoid false readings.

WEATHER COMPENSATION

A type of central, building wide temperature control is weather compensation, whereby the temperature of the water flow from the boiler is regulated by a compensator based on external temperatures. When it's cold outside the water temperature is adjusted to its highest design value for the system it supplies. When the weather becomes milder outside the temperature is reduced preventing overheating and minimising energy consumption.

Boiler controls

BURNER CONTROLS

Most boilers come with a level of burner control whereby the sensors are located in the boiler outlet and modulate the burner to maintain desired temperatures. These should be included within the controls to ensure that they are not operating when there is no demand for heat.

BOILER SEQUENCE CONTROLS

Boiler sequence control should be installed where multiple boilers are used and it regulates boiler switching and modulation to allow only the minimum number of boilers to be turned on to meet the heating demand.

Sequence controls are complex and should be designed and detailed by specialists (mechanical engineers) within the design team.

BOILER PUMP INTERLOCKING CONTROLS

These are necessary as boilers cannot operate unless pumps are running; to extend plant life span boiler pumps should be set to run a couple of minutes past the point where the boiler is turned off to avoid overheating the boiler.

SPACE INTERLOCKING CONTROLS

These are useful for mixed mode buildings and naturally ventilated buildings. They are linked to the opening of windows or doors and switch heating or cooling off if they are left open.

Zoning

Zoning the building with respect to heating, ventilation and air conditioning enhances the effect of control increasing the cost effectiveness of the investment. Zoning allows the environmental conditions in a specific area to be independently controlled to meet the desired conditions. This increases system and building flexibility by controlling parameters such as internal air temperature and ventilation levels in relation to occupancy patterns and orientation.

In buildings without individual zone control, the amount of heating and/or cooling delivered end up being the same throughout the building regardless of how individual areas are performing. This results in energy waste and associated high CO_2 emissions as well as reduced occupant comfort. Splitting the building into a number of separate zones provides closer control to match the desired conditions in each area.

Zoning selection is influenced by:

- Internal heating/cooling requirements – based on the number of occupants, amount of casual heat gains from equipment and other sources of heat as experienced in different areas of the building. For example, corridors and classrooms have different temperature requirements as well as having different casual gains from equipment and occupants;
- Occupancy patterns – separate parts of the building are occupied for different time periods. For example, meeting rooms and open plan office areas have different occupant densities and different occupancy patterns;
- Orientation – more exposed areas need different heating and ventilation requirements than internal areas, similarly southerly facing areas of the building will experience higher solar gains at different times of the day;
- The system serving the area – different controls are needed for different types of system. For example, underfloor heating is controlled very differently to radiators or all air heating systems, it is common for larger buildings to have several different types of system serving different building areas and thus separate zones need to be established for effective control to avoid system clashes and occupant discomfort.

There are often differences between occupant, architectural and control zoning. To ensure good building operation, time will need to be spent so that a collaborative approach is taken to ensure that the needs of each perspective are met and designed in appropriately.

OCCUPANCY

Occupancy sensors are used to ensure that building services operate only when areas are occupied, and systems are turned off when there is nobody present. They are mostly used for lighting systems and some extract ventilation systems such as toilets and storage spaces.

Different sensor types are used for different area types and different operations and they are usually linked to a local timer. For example, absence detection in an open plan office will turn lights off 5 minutes after the area is perceived to be unoccupied.

Power

Before any consideration is given to power generation the design of the building needs be developed and extensively analysed to ensure that it has the lowest possible annual energy consumption requirement. So as discussed at the beginning of this chapter, the building needs to incorporate all applicable and viable passive design solutions possible. For the servicing strategy, energy efficiency measures need to have been exhausted with respect to the fabric performance as well as the equipment chosen to provide the services. Once these steps have been followed and the building is as energy efficient as possible, then the consideration of renewable energy generation solutions is possible.

All power generation solutions require specialist design, detailing, installation and commissioning that is often out with the typical design team skillset. As such this poses a small additional cost and time investment requirement that needs to be considered.

The most common low carbon power generation technologies are touched on in the following sections.

Wind turbines

Building mounted or urban wind turbines are very inefficient and often greatly underperform. This is because the output from a wind turbine is highly sensitive to wind flow and speed. Turbines should be sited away from obstructions, with a clear exposure (often referred to as fetch) for the prevailing wind and this is not possible within an urban or sub-urban environment. The very limited exceptions are when turbines are located on the roof of very tall buildings, where obstructions and surrounding buildings do not interfere with the wind flow. This, however, poses significant issues with the structural loading on the building leading to the need for considerable additional cost to the structure to make any such solution viable. As a solution they are more "for show" than effectiveness and should be considered accordingly.

In rural and heavily exposed areas where the annual wind characteristics are favourable wind turbines are, conversely, very effective. Careful siting with consideration for how future developments may create obstructions is needed. System design considerations include but are not limited to:

- Space, especially important when considering the use of multiple wind turbines;
- Obstructions current and future;
- Wind direction and speed characteristics across at least 1 year.

Photovoltaics (PV)

Photovoltaic solar cells (not to be confused with solar thermal panels) convert solar energy into electricity. The cells are effectively a silicon sandwich with a couple of chemical layers such as phosphorous and boron in between to help charge and help electrons flow to create electricity. It is one of the most effective power generation technologies available, because it has no moving parts and it can be incorporated into most buildings either as a standalone system or as an integral part of the building fabric.

The output from the systems is considered small. In most cases all electricity produced is consumed on site either immediately or with a small delay. PV is thus a supplementary power option and can rarely be considered as the single power source for a building.

The technology is well established and there is a range of options to suit budgets and applications. These include monocrystalline, polycrystalline, thin film and hybrid panels. Hybrid panels are the most expensive and the most energy efficient. On the cheaper end of the scale thin film panels and applications not only produce the least energy per m^2 but also provide for the most adaptable applications. It is primarily thin film PV that are incorporated into shading devices and other building fabric materials such as weatherproof roof membranes or integrated into glazed curtain walling, etc.

The following is a non-exhaustive list of considerations:

- Positioning to maximise the available solar exposure – maximum solar energy conversion occurs when the sun's rays strike perpendicular to the collector surface;
- Orientation to maximise the time the panels are exposed to the sun;
- Overshadowing;
- Cooling – solar panels reduce in conversion efficiency the hotter they get; air should be allowed to circulate freely around the panels to help keep their temperatures down;
- Grid connection and/or battery type and locations to ensure all electricity produced is usable.

Combined heat and power (CHP)

CHP is the on-site generation of electricity and the recovery of the normally wasted heat produced during this process to provide heating. The operation of CHP plant is considered to be a lower carbon technology for producing electricity because it can offer significant CO_2 emission savings per kWh of electricity produced when compared to conventional fossil fuel methods of energy generation. Good quality CHP systems achieve overall efficiencies of 70%–85% by making use of waste heat and eliminating transmission losses.

Generally, CHP is sized and specified based on the heat it produces with electricity typically being less than the total a building will need. If heat is not fully used, then the system effectively acts as a fossil fuel electricity generator and it loses its low CO_2 attractiveness to the degree that electricity becomes cheaper and more carbon efficient if sourced from the national grid.

Notes

1 LEED.
2 This is by no means a complete list. Several publications are available which provide far more detailed guidance on these issues. A good starting point is ISO 13823 – General principles on the design of structures for durability.
3 This list is by no means exhaustive.

4 Efficient design

Consequential savings and costs

Overview

The essential aspect in assessing the cost effectiveness of consequential savings and costs of any given construction alternative is the identification of all the relevant inputs and outputs and quantification, when possible, of these as costs and benefits to facilitate informed decision making. These benefits are more difficult to quantify in monetary terms because they often tend to have more intangibles. Benefits, however, should be as important as costs and deserve to be brought to the attention of decision makers.

Most investment decisions particularly for projects funded by governmental bodies or for industrial applications have a stated goal defined in terms of required or expected output (e.g. kWh/year energy consumed or produced, number of items off a production line, expected occupancy for hotels, pupil performance, recovery rates, etc.). These goals are not always quantified or even quantifiable. However, any form of quantification provides a potential measure of benefits associated with the investment.

A Benefit/Cost Ratio (BCR) should be determined when the output from the investment can be quantified and a uniform annual cost derived from the life-cycle cost analysis. These ratios should be compared for several different alternatives to assist in selection of the most cost-effective options.

Despite best efforts to develop quantitative measures of benefits, there are situations that do not currently lend themselves to such an analysis. A number of actions provide benefits such as improved quality of the working environment, preservation of cultural and historical resources, safety and security of the building occupants and other similar qualitative advantages. Although they are most difficult to assess, these benefits should be documented and portrayed in a life-cycle cost analysis.

Owners of the West Bend Mutual Insurance credited the energy efficient strategies implemented in the new Headquarters Building, West Bend, IN for 99% reduction in personnel complaints about IAQ and 16% improvement in productivity.

For qualitative benefits the more accurate the description the more reduced the potential subjectivity and inherent lack of precision. However, if the following guidelines are observed, qualitative statements can make a positive, effective and important contribution to the analysis.

1 Identify all benefits associated with each alternative under consideration. Give complete details.
2 Identify the benefits common in kind but not to the same degree among the alternatives. Explain all differences in detail.
3 The analytical hierarchy process should be employed. This is one of a set of multiple benefit decision analysis tools that consider non-monetary attributes (qualitative and quantitative) in addition to common economic evaluation measures when evaluating project alternatives. Standard Practice E 1765 Guidelines for Applying the Analytical Hierarchy Process (AHP) to Multi-Attribute Decision Analysis of Investments Related to Buildings and Building System published by ASTM International presents a procedure for calculating and interpreting AHP scores of a project's total overall desirability when making building-related capital investment decisions.

Following these general guidelines will help document these intangibles that are measured in non-economic terms like aesthetics, safety or morale, and enhance the value of benefit/cost analyses and make informed decision-making easier.

It is also noted that in addition to benefits, information concerning negative aspects of alternatives, quantified where possible, should also be included to ensure the objectivity and completeness of the analysis. This information is important in decision making and possibly to the community at large and may be a determining factor in deciding between possible investment alternatives.

Externalities are an important class of outputs that may be benefits or disadvantages. Air pollution is an example of an externality that is not a benefit.

We are trying to set a standard and recommended format for qualitative benefit analysis, with the importance on the content. No analysis is truly complete unless it addresses benefits attending all the alternatives under consideration.

Secondary and tertiary benefits to be assessed

Occupant wellbeing and productivity – comfort

It is notoriously hard to accurately and consistently measure human productivity. Studies that do offer some information are not considered robust enough with respect to reporting on the differences between buildings, their management and their operation. Additionally, our behaviour is heavily influenced by factors not at all linked to building conditions, such as whether we are in a group or in isolation.

Despite these difficulties, over 20 years of data from detailed studies and questionnaire data from post-occupancy building studies have shown differences in productivity to varying degrees based on different aspects of comfort. The absence of productivity is more measurable in these cases, as is the lack of comfort. In essence, it is extremely difficult to quantify the effects of buildings working well, but in recent years we have been able to quantify failure, and more specifically quantify it not only in terms of maintenance, or energy costs but in productivity.

Unsurprisingly, perhaps to those of us with experience of hundreds of different buildings and their workings, the data shows strong correlations between increased comfort and thus productivity and those buildings characterised by design, construction and operational practises in line with the principles of energy efficiency and environmental friendliness. Consequentially, the further back we go to look for data, the less of it there is, as buildings tended not to be designed with those aspects in mind; instead the existing building stock is full of unmanageably complex buildings that underperform. Thus, we have to rely on relatively new data compiled from comparisons between older buildings with the limited number of newer ones available; nonetheless, the results are compelling.

Overall comfort is affected by numerous factors, and yet until recently buildings have been designed with checks on comfort solely relying on internal temperature performance. To a great degree this is because temperature is perceived to be the most "controllable" aspect. Regardless, the premise is flawed as are consequently the results. Comfort is determined by many more aspects that can be influenced by design and building control. And these aspects, by extension, influence productivity:

- surfaces,
- volumes,
- glazing,
- radiant and air temperatures,
- humidity,
- air movement,
- lighting,
- clothing,
- activity,
- age,
- gender,
- fitness level,
- health,
- time of day.

However, the first seven aspects are down to design decisions alone, and they do make a measurable difference. The following explains how comfort is affected by these and how it impacts on productivity.

Air quality

The quality of the air we breathe is paramount. Taking the general view, the health effects of air pollution are well documented and increasingly more publicised through media. It is not only exposure to carcinogens like asbestos fibres or passive smoking that need to be addressed. There is solid evidence to link mortality to exposure to particulate matter in the air we breathe amounting to £100s of millions per year per city.

Beyond mortality, expensive, long term health problems such as asthma have been on the rise for decades and their effects on businesses are equally significant. Simple, run-of-the-mill, filtering of incoming air is clearly not cutting it. If it were, the numbers would not be increasing, bearing in mind that the vast majority of people spend their time indoors, the filtration systems in use would have stemmed the tide, and they have not. Research in fact shows that approximately 65% of outdoor air particle inhalation occurs while indoors and so the adverse effects of air pollution are still present when indoors unless care is taken in the design of the ventilation systems.

The main pollutants are:

- CO_2 from human respiration within the space;
- Fumes and air borne contaminants – such as particulate matter (PM10 and PM2.5) gaseous pollutants such as CO, NO_2, SO_2 ozone and VOCs that are not filtered out of the outside air via ventilation systems;
- Tobacco smoke – although smoking indoors is prohibited by law in most countries, second-hand smoke can still enter a non-domestic building;
- Emissions from building materials, furnishings and equipment such as ozone from laser jet printers, photocopiers, ultraviolet lighting and, formaldehyde from glues and adhesives used in furniture and finishes, and, off gassing from carpets and soft furnishings;
- Bacteria and dust spread by inadequately filtered and poorly maintained heating, ventilation and air conditioning (HVAC) systems;
- Pesticides sprayed on plants;
- Carbon monoxide produced from gas and paraffin burners;
- Irritants – from pets, dust mites, cockroaches and some plants;
- Bacteria – from excessively humid environments, mould in damp areas and rotting food;

- Radon gas – entering the building through the foundations and air intakes.

A more detailed approach is needed taking into account all aspects contributing to the quality of the air we breathe.

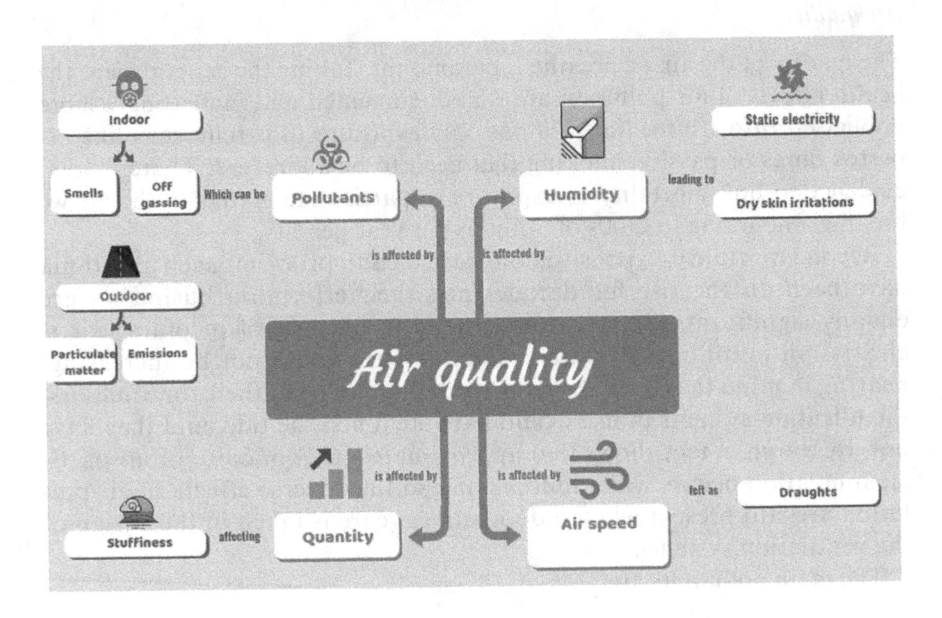

There are many design decisions that affect air quality. The design for good air quality is straightforward, as many sources of pollution as possible need to be identified and a good quality environmental control system needs to be provided to ensure that these pollutants are consistently removed.

When considering the BCR for the following aspects of design, the impact on air quality needs also to be considered:

- Building location with respect to external sources of pollution – this will determine the quality of the external air and whether natural ventilation is a potential option for the building servicing strategies;
- Building shape – this dictates the type of ventilation strategy that can be applied, long spans and low ceilings are not conducive to natural ventilation;
- The extent and thoroughness of site contamination remediation – this will determine whether any residual gases are likely to enter the building after completion (such as radon);
- Building airtightness – this will dictate the air volumes required for good purging of internal pollutants;

- Material choices for structure – this will determine whether residual contamination is likely in the form of interstitial condensation leading to mould growth;
- Material choices for finishes, fittings and furnishings – this will reduce the emission of volatile organic compounds such as formaldehyde and toluene;
- Material detailing with respect to paints and adhesives – this will reduce the emission of VOCs, paint toxins, adhesive toxins, mercury and lead;
- Building ventilation strategy;
  - Location of air inlets (natural ventilation openings or mechanical ventilation air intakes) – location determined based on convenience alone can compromise the quality of air coming in; for example, locating openings at street level introduces pollutants such as particular matter and second hand tobacco smoke from people outside;
  - Mechanical ventilation system design air volume levels – higher ventilation volumes beyond minimum regulation compliance levels provide better air quality as contaminants are not allowed to concentrate to noticeable levels. Many studies have shown that the percentage of occupants dissatisfied with their indoor environment reduced in line with the increase in ventilation rates;
  - Mechanical ventilation system specification for filtration – different types of filter provide different levels of filtration and in some cases multiple types may be needed to provide good air quality; Class G4 and F7 filters help with particulate matter and activated carbon filters help with gaseous pollutants;
  - System maintainability – regardless of the quality of the system the ability to easily and quickly maintain it is as important as its specification and construction; if access to systems that need replacement such as filters is not adequately provided, then their effects are as good as their life span which in some cases can be as short as 3 months.

A compromise on any of these aspects will compromise overall air quality, which will compromise productivity and the perception of building quality. This will mean consequential losses for the building owners and its occupants. To fully appreciate the BCR thus these potential losses need to be accounted for and brought out in the open to enable appropriate decisions to be taken. The most important time to do this is at the value engineering exercises undertaken towards the end of the design cycle, where cost-cutting discussions often disregard the full design intent and lead to decisions that result in subpar building performance.

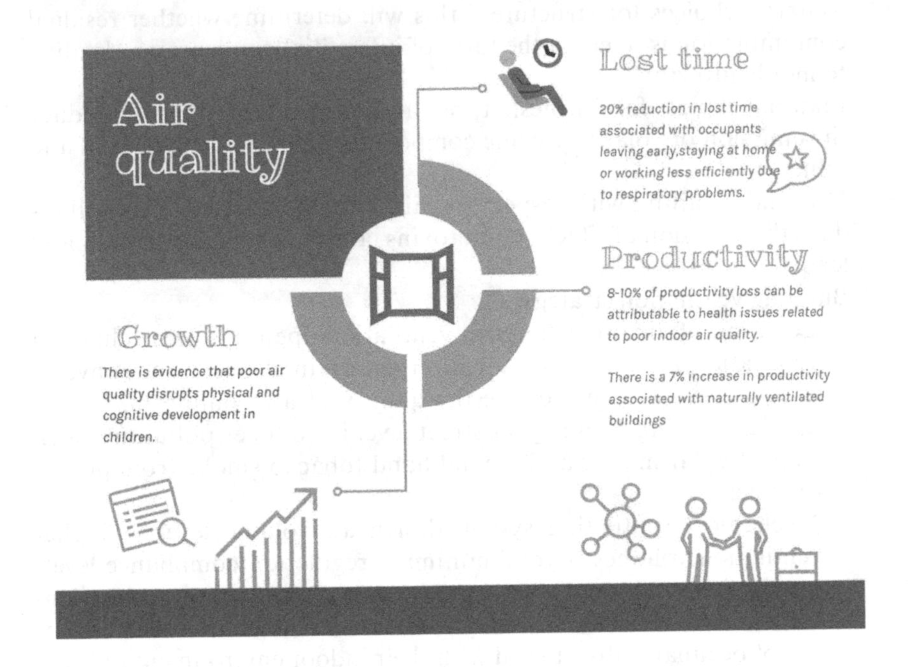

Visual comfort

The health effects of bad lighting design may not be as severe as those due to poor air quality; they are nevertheless present. The impacts of continued exposure to poor lighting conditions on our health include migraines, eye strain, fatigue and sluggishness, stress and depression.

In particular, glare is associated with visual discomfort and eye fatigue as well as headaches and migraines, and this can lead to accidents. Access to daylight, or the lack thereof, affects our circadian system, and this governs digestion, the release of certain hormones, body temperature and sleep. However, a badly functioning circadian system can also lead to obesity, diabetes, depression and other metabolic disorders; in fact, there has also been a link between certain types of cancer and circadian phase disruption. Furthermore, irregular sleep-wake cycles are closely linked to poorer performance.

In contrast, the provision of a high-quality visual environment enhances productivity and helps maintain healthy circadian rhythms. The primary characteristic of increased visual comfort is the availability of daylight; the rest of the main basic principles of visual quality are as follows:

- Visual performance – When people can see the task better, they can perform better;
- Visual comfort – Decreasing discomfort increases concentrations and improves performance;
- Visual ambience – Lighting influences visual ambience, which in turn, influences performance;

- Interpersonal relationships – Lighting influences not only how we perceive objects but also how we see each other; this influences co-operation and by extension productivity. For example, it was found in 1992 that lighting conditions influence the likelihood that occupants would choose co-operative rather than competitive forms of conflict resolution;
- Biological clock – Light adjusts our biological clock, which controls our circadian rhythms and thus influences performance at certain times of the day as well as influencing the quality of our sleep;
- Stimulation – Light stimulates psychological and physiological processes, which enhance performance. An example is an increase in job satisfaction by way of improved task significance and autonomy.

Designing for a good visual environment is well established with significant publications available on the details required for best practice. Impacts on visual comfort should be considered when assessing the BCR for the following design aspects:

- Building orientation – this will dictate how the building interacts with the angle of the sun across the year;
- Building shape – this constrains the type of lighting design, deep plans and low ceiling heights may offer some capital cost savings, but they are not at all conducive to good quality lighting design;
- Useful daylight index calculations and daylight availability – these will allow an understanding of how openings affect the quality and distribution of daylight within the space;
- Glare management and calculations – these are necessary to increase lighting quality and reduce the distractions, because glare within peripheral vision leads to increased looking away from the tasks at hand reducing productivity and increasing eye strain;
- The extent of lighting and daylighting calculations – it is not sufficient to just "double-check" hand calculations with a final simulation check if visual comfort is a priority, calculations and time to complete them as well as an allowance for changes based on results need to be incorporated as an item that adds value to the final building.
- Glazing location – this will need to be dictated by the UDI as well as ventilation assessments;
- Glazing choice – the coatings and colour of the glass chosen will impact on daylight quality and the thermal comfort of the occupants;
- Light levels – light levels need to be appropriate for the tasks carried out, with clear strategies for general background lighting and task specific lighting appropriate to the visual acuity required;
- Light distribution, surface colours and illuminance – areas can look gloomy and constricted if light is not distributed to the ceiling, walls and floor appropriately;
- Colour rendering and colour temperature detailing – our circadian response is dependent not only on the light that enters the eye but also on factors such as spectral properties of the light, brightness levels, duration and timing of exposure;
- Surface reflectance specification and detailing – glare off surfaces can be as distracting as glare from the sun;
- Lighting control – the ability to control light levels and direction provides a minimum 10% improvement on the perception of good visual quality.

Acoustic comfort

The difference between sounds and noise is slightly subjective but often it is influenced by the function of the space we are in. In general terms, noise is unwanted sound that may be unpleasant and that causes disturbance.

Taking this one step further, noise nuisance is considered excessive noise or disturbance that may have a negative effect on health or quality of life. The term excessive refers to length of exposure to the noise and/or the level of the noise. The higher the level of noise, and the longer individuals are exposed to it, the more risk they have of suffering harm from it.

However, not just loud or sudden noises provoke a stress response. Chronic low-level noise also negatively influences our brain and behaviour. Whether from external or internal, low-intensity noise has a subtle yet insidious effect on our health and well-being. For school children, noise at home or school can affect a child's ability to learn. Compared to children from quieter neighbourhoods, those living near airports or busy highways have measurably lower reading scores and develop language skills more slowly. This presents slightly differently for adults whereby workers in noisy offices experience significantly higher levels of stress (as measured by urinary epinephrine, a stress hormone), and in a study, they made 40% fewer attempts to solve an unsolvable puzzle and made only half as many ergonomic adjustments to their workstations, compared to their colleagues in quiet offices. Additionally, it should be noted that psychiatric hospitalisations are higher in noisy communities.

Bad moods, lack of concentration, fatigue and poor work performance can result from continual exposure to unpleasant noise. As such, productivity is significantly impacted upon when there are distractions from internal or external noise. Although distraction can often be affected by an individual's propensity, there is considerable evidence to show that acoustic comfort is important; so much so, that there are legislative limits associated with noise levels in almost all country specific building codes.

The UK government defines three levels of noise: the level of noise exposure where there is no effect on health or quality of life; the level of noise that has detectable adverse effects; and the level of noise exposure where there are significant effects on health and quality of life.

Noise-induced hearing loss is the most common preventable occupational health condition in the world. Interestingly, research suggests that it takes 10 years from the time someone notices they have some hearing loss before they do anything about it. Taking into account that one in six (10 million) individuals in the UK have with some degree of hearing impairment, increasing acoustic comfort in buildings is not only logical but can provide significant benefits with respect to productivity increase. Looking at some of the common signs of hearing impairment we can see a clear correlation between these symptoms and low productivity:

- regularly feeling tired or stressed, from having to concentrate while listening;
- trouble understanding conversations at a distance or in a crowd;
- finding it difficult to tell which direction noise is coming from;
- answering or responding inappropriately in conversations;
- feeling annoyed at other people because of not understanding them;

- feeling nervous about trying to hear and understand others;
- reading lips or more intently watching people's faces during conversations;
- ringing in the ears;
- inability to hear soft and high-pitched sounds;
- muffling of speech and other sounds.

Not only can this issue reduce an individual's efficiency it can also lead to accidents due to limited speech communication, misunderstanding oral instructions and masking the sounds of approaching danger or warnings.

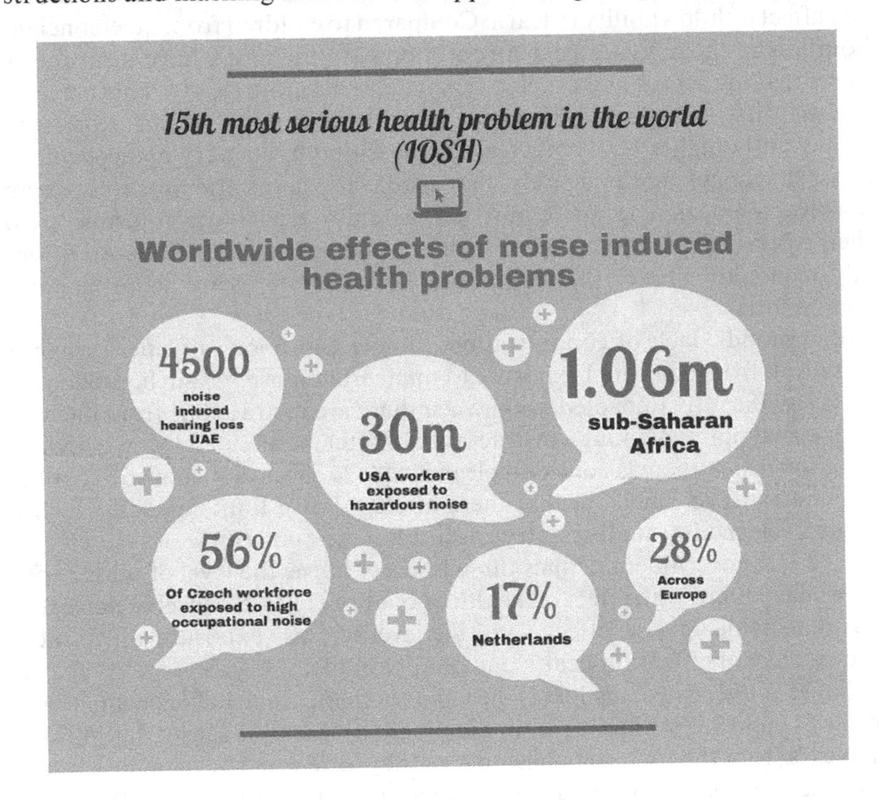

In Europe alone the WHO estimates the annual cost associated with the health effects of excessive noise to be around £30 billion. Hearing impairment is the most well-known effect of prolonged aural discomfort; however, noise can also exacerbate a multitude of health conditions:

- Stress – Persistent noise in the work environment can be a stressor even at quite low levels. In a study in Austria the children in noisier neighbourhoods experienced higher overnight levels of the stress hormone cortisol, marginally higher resting systolic blood pressure and greater heart rate reactivity to a stress test – all signs of modestly elevated physiological stress;

- Tinnitus – (ringing in the ears) is the early sign of hearing damage. It can be a very distressing condition and can lead to disturbed sleep and affected speech. Furthermore, there is no effective cure, it is permanent;
- Noise-induced hearing loss – is the most common preventable occupational health condition across the world. The level of noise that is likely to damage hearing varies depending on the individual's characteristics and the duration for which they're exposed to the noise. Acoustic trauma also falls under this category. Once hearing has been damaged, there's currently no known cure or effective treatment;
- Effect on pregnancy – Exposure of pregnant individuals to high noise levels can affect the unborn child. Research suggests that prolonged exposure of the unborn child to high noise levels during pregnancy may have an effect on a child's later hearing and that low frequencies have a greater potential for causing harm;
- Physiological effects – Noise can have an effect on the cardiovascular system, resulting in an increase in blood pressure and the release of catecholamines in the blood. An increased level of catecholamines in the blood is associated with stress.

These effects are not only applicable to "high risk" or noisy scenarios; unfortunately, the non-auditory effects of noise appear to occur at levels far below those required to damage hearing and thus, distraction from noise is often one of the lead causes of dissatisfaction with the office environment for the same reasons. Even though the distraction depends upon the task at hand, the acoustic environment and the individual's cognitive characteristics, excessive discernible noise is potentially responsible for greater dissatisfaction and productivity loss than any other single environmental factor. A world green building council report (WGBC) found that there was a 66% drop in performance from people exposed to distracting noise. This is a statistically significant finding especially when these workers operated in environments that were designed to legislative requirements for noise reduction. A survey in 2005 identified that 99% of people reported that their concentration was impaired by office noise such as unanswered phones and background speech.

Thus, acoustic comfort should be a factor to consider when assessing the BCR for the following design aspects:

- Building location – the degree of external noise associated with a building's location should be one of the considerations for choosing a site;
- Landscaping – green landscaping if appropriately assessed and detailed can play a significant role in softening noisy external environments;
- Building materials – in particular, sound insulation and the effects of material detailing on internal reverberation times;

- Furniture finishes and fittings – more reverberant acoustics can also result in aural discomfort, so a balance has to be reached between material choices and the incorporation of sound absorbing materials into fittings and fixtures;
- Space layout and size – can affect how sound behaves and careful consideration should be given on the effect of the geometry of the space on the Speech Transmission Index (STI);
- Ventilation strategy – a common problem with buildings that rely on opening windows for ventilation is the ingress of external noise and this may reduce background sound to below target levels, at which point it may be desirable to add background sound to help raise it to an acceptable level for masking distracting noises.

Thermal comfort

Thermal comfort is the result of the workings of a self-regulating complex and adaptive system. To assess thermal comfort based on temperature alone is not only misleading but also pointless. In reality, thermal comfort is influenced by radiant and air temperatures, humidity, air movement, clothing, activity, age, gender, fitness level, health, time of day. Achieving and maintaining thermal comfort for occupants is one of the primary aims of building design and yet the assessment of whether a building is inherently comfortable or not (a free floating thermal comfort analysis) has been flawed for many years. This impacts greatly on the way buildings need to be designed and operated. Our adaptivity to the environment affects the type of plant and amount of energy required to heat and cool our buildings. The available standards which provide prescribed "ideal conditions" for thermal comfort have been historically based on the use of centralised mechanical systems providing little guidance on how such narrow comfort bands can be applied to naturally ventilated buildings whose conditions are much more variable.

It is simplistic to disregard the fact that occupants play an instrumental role in creating their own thermal preferences through the ways they are allowed to interact with their environment or that they gradually adapt their expectations to match the thermal environment they are in. For over 40 years research has shown that allowing people greater control over their own indoor environment and allowing indoor temperatures to more closely track outdoor climatic patterns have significant positive effects on improving comfort. In other words, the traditional assessment of thermal comfort based on narrow air temperature bands cannot be an accurate determinant of whether a building needs mechanical ventilation and refrigerant cooling. Although on the base of it setting an upper limit on indoor air temperature is fundamentally sound, the assumption that there is a single indoor temperature limit irrespective of outdoor conditions is no longer considered sufficient.[1]

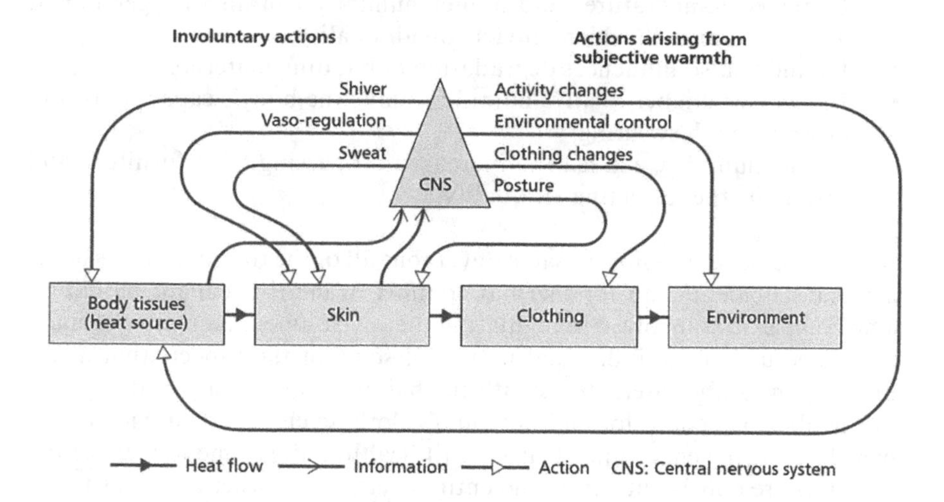

Newer, more detailed and complex assessments that take into account the adaptive nature of our thermoregulating system need to be employed to make accurate decisions on the performance of the building.

This is important because the indoor thermal environment not only impacts on approximately half of a building's energy consumption, it also plays a large role in the way we experience the building itself with respect to our pleasure in the building. Thermal comfort is so closely linked to occupant health, well-being and productivity that it is consistently ranked as one of the top three contributing factors influencing overall occupant satisfaction in buildings. Our thermal experience in a building influences the body's integumentary, endocrine and respiratory systems so closely that in some cases the effect of a thermal discomfort can be instantaneous. For example:

- Exposure to cold air and sudden temperature change can rapidly exacerbate asthma symptoms in adults. Link this with low air quality and the situation can become a perfect storm for some;
- Cold spaces with low relative humidity facilitate the spread of the influenza virus as dryness allows the virus to persist longer in the air and lower air temperatures help extend virus shedding periods;
- Lower relative humidity in winter can lead to dryness and irritation of the airways, skin, eyes, throat and mucous membranes;
- Overheated spaces are linked to increases in sick building syndrome symptoms which include irregular heart rate, respiratory issues, fatigue and negative mood;

- Increased temperatures and higher humidity can also trigger mould and fungal growth which impacts on air quality;
- Humidity also influences degradation of building materials;
- In summer, higher relative humidity limits the body's capacity to cool down through sweating;
- Higher humidity can lead to increased off-gassing from furniture and fittings further affecting air quality.

It is, of course, impossible to satisfy everyone all of the time and this is never a truer statement than for thermal comfort in multi-occupant buildings. The complexities involved in regulating the environment increase exponentially because of the individual nature of some of the aspects that affect our responses and interactions with the buildings we occupy. Namely, it is impossible to account for an individual's clothing choice, their age and fitness level their gender and their overall health and assume that these parameters are consistent across the entire occupancy. Other aspects such as thermal history and climatological origin, temperament, preferences, social and cultural expectations and seasonal variation also influence an individual's experience of thermal comfort. It is understandable that designers and stakeholders seem to steer clear of looking too deeply into thermal comfort beyond just temperature.

Extreme examples of this are found in some building types such as hospitals and nursing homes, etc. whereby design and controls may be able to accommodate for the patients but leave the staff (who invariably have different individual requirements) uncomfortable; that means a significant percentage of occupants are not in an optimal thermal environment. Some accommodations can be made, such as provision of areas or zones kept at different conditions to help provide respite to the uncomfortable for parts of the day. Even in non-specialist buildings one size certainly does not fit all. Someone who drives to a building and parks their car within a few yards will experience the building in a completely different way than someone who has walked or cycled to it.

It is a difficult aspect to get right but there are techniques and calculations that can make it easier to provide a base level of comfort to accommodate around 80% of occupants. Encouragingly, studies have shown that within certain temperature ranges (say between 16°C and 24°C) occupants are more adaptable to temperature in a way that they are not, to air quality. Flexibility is the key to decreasing the percentage of occupants that experience prolonged discomfort. In essence, the design should enable occupants to make local system adjustments easily to allow for individual thermal preferences. This sense of control helps adjust expectations to the degree that people that would otherwise consider themselves to be uncomfortable, would be more inclined to accept slightly discomforting conditions without comment. Sub-zones should be incorporated that allow for this.

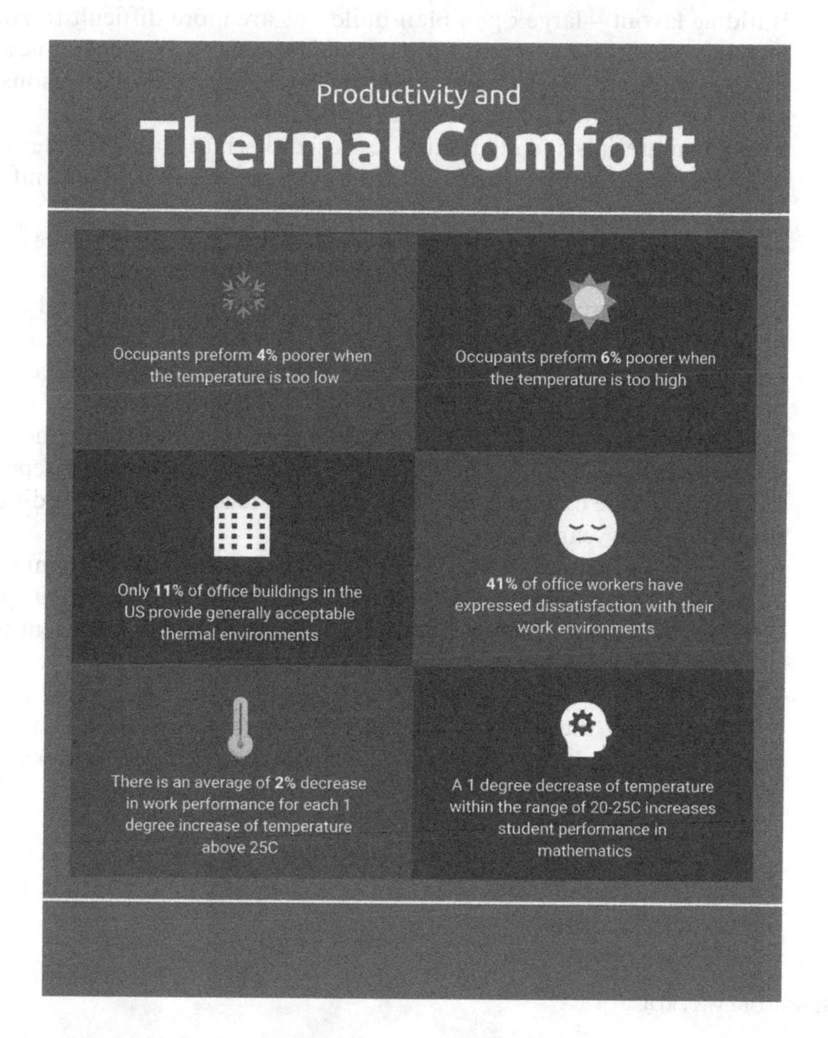

An "adaptive" comfort strategy allows temperatures to drift down in winter and up in summer and helps mixed mode ventilation become viable for more buildings. However, the benefits this sort of strategy can only be realised if users must have effective control of their environment, and this is why such a model works exceptionally well in domestic buildings but falls apart in some commercial applications.

It is clear that thermal comfort should be a major contributing factor to consider when assessing the BCR for the following design aspects:

- Building shape – dictates to a great degree the heating ventilation and air conditioning choices viable for servicing the building limiting the opportunities for flexibility and adaptive thermal comfort strategies;

- Building layout – large open plan buildings are more difficult to zone, providing layers between the outdoor and indoors such as curtains, and shutters offer more options to adapt the building across the seasons to create more comfortable internal conditions;
- Building materials and structure – exposed thermal mass provides regulation to high temperature swings allowing occupants to stay comfortable during external extremes on a day-to-day basis;
- Air movement – increasing draughts in summer provides effective cooling without the need for refrigerants;
- Landscaping – green landscaping can reduce the local heat island effect providing the opportunity to reduce extreme temperatures in the microclimate around a building reducing the need for refrigerant cooling allowing for greater adaptability;
- Servicing strategy – the servicing strategy will dictate what thermal comfort strategy can be employed in the building and highly dependent on this is the cost of effective controls to reach comfort conditions throughout the year;
  - Providing attention in the design to the active control of radiant temperatures from systems like heating radiators or chilled ceilings provides the benefit of better thermal comfort and a more efficient way of generating and transporting cooling.
- Thermal comfort simulation – the investment in time and effort to accurately predict thermal comfort for occupants will influence building performance. Multiple simulations should be carried out to assess the effect of all potential solutions at all detailed design stages as well at concept stages;
- Control strategy – the impact of controllability on occupant comfort cannot be underestimated. A sense of control is paramount to the effectiveness of any thermal comfort strategy.

SENSE OF CONTROL

The provision of a sense of control may not seem immediately work-related, but the ability to create and control one's entire day as the needs of work dictate puts employees squarely in control of their own time management, productivity and processes. This has been shown to lead to greater organisational productivity and suggests that meeting an employee's need for autonomy influences motivation and performance. In fact, a 1980s study into sick building syndrome, or building related sickness as its now known, confirmed that occupants' perception of control over their environment affects their comfort and overall satisfaction as their tolerance for conditions increases. This is substantiated by numerous studies that indicate that lack of control is people's single most important concern.

In building design and control terms, this effectively means that, if an occupant has more control over their environment, they tend to be more satisfied as a result. A study found that relatively limited individual control over temperature led to an increase of about 3% in logical thinking performance and 7% in typing performance. This was corroborated by a separate study that found 3% gains in overall productivity as a result of personal control of workspace temperature.

The benefits of increasing the sense of control can be realised even beyond just temperature. Control of air supply and speed (for example, from desk mounted fans) improves the perception of air quality. Control of light levels improves satisfaction and mood. These then influence motivation task performance to this list of benefits.

Interestingly, a multilayer study was carried out to investigate how having or not having control over one's indoor climate affects work performance in office buildings. The study showed that occupants that have a high amount of control over their indoor climate perceive themselves to be significantly more productive than those that have a low amount of control. The quantitative effect of improving a no control situation towards a full control situation was estimated to be at least 6%.

It should be noted that the building industry cannot be immune to technological advances and the rate of development of wireless sensor technology, which enables sophisticated monitoring and logging of temperature, humidity and lighting should be acknowledged. Other innovations include building control apps that allow people in commercial buildings to "vote" and influence the operation of building services without use of thermostats or intervention by building operators. Bearing in mind that digital technology is likely to become embedded more and more in building structures and equipment as it is already becoming embedded in terms of wearable technology such as smart watches.

Control systems should be assessed on the following criteria beyond just cost:

- High levels of lighting system, ventilation and thermal comfort control for individual workstations to promote the productivity, comfort and well-being of building occupants;
- The ability for zoning in multi-occupant spaces for heating, cooling, lighting, ventilation, noise and privacy (e.g., classrooms or conference areas);
- How they support the effective operations and maintenance of building systems so that they continue to meet target building performance requirements over the long term;
- How many unnecessary functions are assigned to automatic control;
- How clear user interfaccs and labels are;
- How responsive they are.

Mental health

Health is not simply determined by the absence of a condition, it is an overall state in which individuals are able to live to their fullest potential, cope with the normal stresses of life, work productively and contribute to their community. Mental health is as fundamental a component to this as physical health is. Of course, it is an extremely complex aspect that is determined by a range of socioeconomic, biological and environmental factors, such as age, gender, working and living conditions, lifestyle choices, health behaviours and genetics all of which influence chemicals in the brain.

The effects of mental ill health are so significant that it is beginning to be considered a global epidemic:

- Mental health, and subsequent substance misuse, conditions collectively account for 13% of the global burden of disease (GBD) and an estimated 32% of years lived with a disability;
  - Alcohol alone accounts for 3.3 million deaths per year (or 6% of all deaths) and 5% of GBD;
  - Depression alone accounts for 4% of GBD and is among the largest causes of disability worldwide;
  - In total, it is estimated that over 14% of deaths worldwide are attributable to mental health conditions.
- A staggering 30% of adults will experience a mental health condition during their lifetime and 2/3rds of these adults are employed in the workforce;
  - In high-income countries, 35%–50% of these people receive no care or treatment;
  - In low- and middle-income countries it is up to 85%.
- When left unmanaged, mental health conditions – especially depression – can place an individual at risk for suicidal thoughts, attempted suicide and completed suicide.
- Depression increases the risk of disease such as, diabetes, cancer, cardiovascular disease and asthma;
- States of chronic stress increase the risk of conditions such as depression, cardiovascular disease, diabetes and upper respiratory infection.
- People with mental health conditions experience a mortality rate 2.2 times higher than the general population and a median of 10 years of potential life lost.

Increasingly, research into mental health indicates that building design can play a significant role in addressing some of these negative aspects and in supporting some of the drivers for good mental health. For example, the biophilia hypothesis states that humans have an innate desire, to seek connection with nature and other forms of life; this is an integral part of human biology and evolution and as such, it is fundamental to human health and emotional wellbeing. Biophilia translates as the love/passion (philia) of life (bio) or living systems. It was used by a social psychologist named Erich Fromm (1973), who used it to describe the "passionate love of life and of all that is alive".

The effects of our lack of this connection to nature have become ever more prevalent in more recent years. The way in which buildings have been designed leave them often devoid of nature and other forms of life. The prevalent practice of fully automating the control in our buildings (and oftentimes doing it ineffectively) frequently leave occupants lacking in a sense of contribution, engagement or ownership of their space and with their environment. This can manifest in increased mental health disorders such as stress, anxiety and depression. Even on a neighbourhood level, in our cities and conurbations, it can be challenging to find enough opportunities to satisfy our innate need to affiliate with and positively contribute to natural systems and processes.

Conversely, multiple peer reviewed studies have shown that a connection with nature improves stress recovery rates, cognitive function, enhances mental stamina and focus, decreases violence, elevates moods and increases learning rates.

Designing with biophilia in mind helps bring this connection to nature to our internal environments and can include many features such as views out to a natural landscape, indoor plants and greenery, natural light and systems linked to our circadian rhythms, natural tones and textures, natural materials, sounds of nature, water features or even natural shapes and forms (biomimicry[2]).

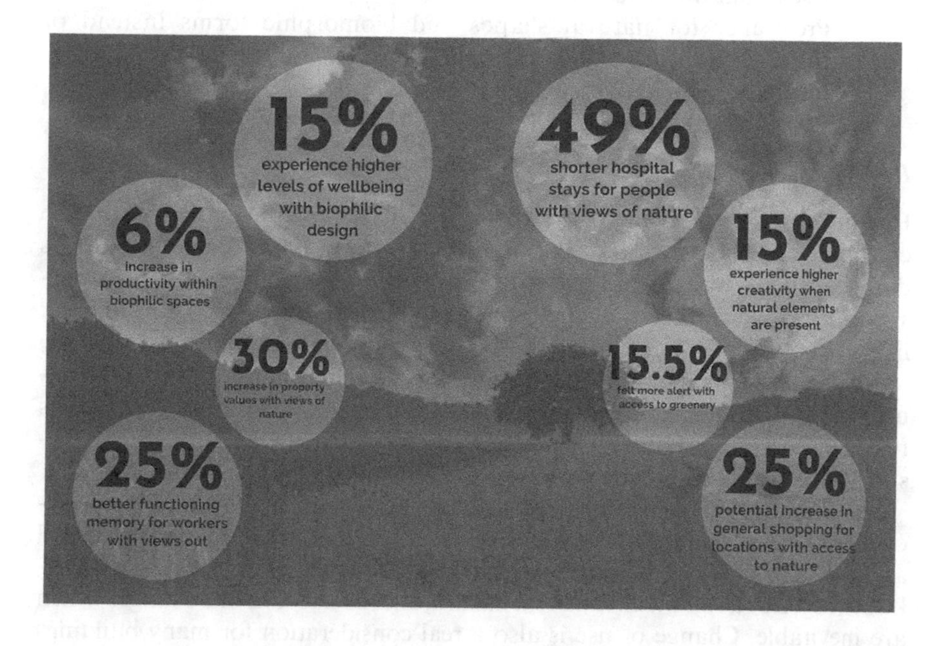

Occupant wellbeing in general is a strong selling point even for developers looking to capitalise quickly on their building development investment. Any building designed and constructed with human comfort at its centre presents an excellent and powerful differentiator within a market inundated with sub-standard offerings. Beyond that, owner occupiers stand to benefit consistently over many years from buildings that provide true enjoyment to users. Well-being and mental health come from a combination of air quality, visual and thermal comfort and effective and efficient good building control. As such all design aspects discussed to this point affect the quality of the building.

The main points to consider when assessing the BCR with respect to mental health are:

- Building shape and layout – dictates access to views out for occupants;
- Building materials and structure – dictate the level of biophilia the structure and building envelope may provide;
- Building location and landscaping – influence the access to and views of nature available;
- Servicing and control strategies – influence air quality and visual comfort and the sense of control occupants will have over their environment;
- Furniture, fixtures and finishes – provide the opportunity to enhance biophilia within the building through practices such as:
  - Use of natural materials as timber, stone and wool instead of synthetic materials;
  - Incorporation of pictures and artwork that depict nature;
  - Use of natural colour palettes;
  - Preference for natural shapes and biomorphic forms instead of straight lines;
  - Inclusion of or provisions for indoor plants.

Resilience

Resilience is determined by the building's ability to adapt and respond to changing conditions while still remaining functional. It is a good building's defining characteristic and it is affected by how long the building can consistently provide for basic occupant needs: potable water; sanitation; energy; liveable conditions; light; safe air; and occupant health.

Resilience is not absolute and complete resilience against all potential situations is, of course, not possible. The idea is to make incremental decisions to accommodate feasible initiatives in order to achieve a greater overall result which withstands the impacts of the most likely risks.

This can be done when these potential hazards are identified and the development's vulnerabilities to them are assessed. The standard hazards that will affect all buildings are considered to be climate change which is globally acknowledged as a significant and real hazard; and, socioeconomic changes which are inevitable. Change of use is also a real consideration for many buildings developed to last over 60 years. Essentially, adaptation to a changing climate which may mean higher temperatures, more intense storms, sea level rise, flooding, drought and wildfire is a growing necessity, while non-climate-related natural disasters, such as earthquakes and solar flares, and anthropogenic actions like terrorism and cyberterrorism, also call for resilient design.

The decision on resilient design, beyond what is required by legislation, is one that has to be taken at early stages, and potential scenarios and impacts need to be decided upon, before any other decisions, as they will affect how and what performance targets are established which, in turn, will dictate how all further decisions are made starting with site selection. Additional investment may be required to allow for additional servicing or for upgrading the specification of existing designs to meet higher climate thresholds (such as raising road levels or increasing drainage). As such it will require dedicated assessments and decisions. These resilience decisions should be discussed and incorporated during the predevelopment and planning stages of the development.

During the design process a building's resilience can be increased by refining the building design of measures such as:

- Choosing structures that can accommodate the anticipated impacts of change;
- Selecting materials and components that will not present a hazard in the event of damage;
- Identifying and strengthening such critical systems to withstand extreme weather;
- Using future climatic conditions to model design solutions;
- Designing for habitability in the event of power or energy loss;
- Incorporating simple, passive and flexible systems that are able to adapt to changing conditions both in the short- and long-term;
- Building in diversity and designing in layers to allow for partial breakdown/failure without compromising the entire building;
- Incorporate strategies that protect the natural environment to enhance resilience for all living systems;
- Diversify energy production by optimising on-site renewable energy supply;
- Carrying out more detailed stress testing of design solutions with the identified resilience hazards in mind;
- Reducing the urban heat island effect;
- Designing for sustainable urban drainage systems and to allow for managing stormwater as well as maintaining and protecting aquifers.

As with all consequential savings calculations, estimating the benefits of investing in increasing building resilience is difficult. However, the benefits are real and many organisations are now looking to put a price on them. For example, the Asian Development Bank's (ADB) Urban Climate Change Resilience Trust Fund (UCCRTF) which is supported by the Rockefeller Foundation and the Swiss and UK governments is valuing the urban resilience benefits of the bank's infrastructure loans and technical assistance programs.

In the US, the National Oceanic and Atmospheric Administration (NOAA) declared 2017 the costliest year on record for weather and climate disasters. It identified that mitigation investment can save the nation $6 in future disaster costs; for every $1 spent on hazard mitigation this is a BCR of 6:1. When considering the BCR for investing in mitigation measures beyond simple legislative compliance have can save the nation $4 for every $1 spent (BCR 4:1). The kind of measures assessed included flood resistance, wind resistance, earthquake resistance and fire resistance.

Other studies that have valued the costs and benefits of investing in resilience with respect climate change adaptation and disaster risk reduction generally report positive BCRs with economic returns usually at least three times those of the original investment (3:1) and some projects delivering BCRs of up to 50:1. Nonetheless, building for resilience often requires new investment compared to a business-as-usual scenario and this should be taken into account. It should also be noted that resilience benefits are likely to arise long after the project has been completed and will accrue over the building's lifetime potentially up to 60 years or more.

There are significant benefits to designing for resilience and these should be considered for each project. For example, city planning and construction legislative requirements do not always offer adequate resilience, so going beyond these is necessary. When this is done and, of course, communicated the building becomes significantly more attractive to tenants, buyers and investors and much cheaper and easier to insure.

Additionally, with the likelihood of extreme weather on the increase resilience will generate value in much shorter terms than it has historically. Conversely lack of resilience will mean the building life span will decrease and the reputational risk alone that poses to the developer should be considered carefully.

Conventional cost-benefit analyses of resilience measures are heavily loss-centric. They tend to weigh up the certain cost of investing in resilience against the uncertain benefit of avoided losses and this can result in decision makers underestimating the benefits. Indeed, increasing resilience reduces the background risk of disasters; however, as a consequence, individuals and organisations can be encouraged to become less risk averse. This means they will tend to undertake long term investments in productive assets, engage in entrepreneurial activities and lengthen their planning horizons, all of which are needed for economic development and growth. This is crucial if we are to take pro-actively advantage of structural changes in the global economy, new market opportunities and develop new market niches.

There are increasingly more tools available to help with the assessment of BCR with respect to resilience. For example, GRESB which assesses the Environmental, Social and Governance (ESG) performance of real estate and infrastructure portfolios and assets worldwide has developed a resilience module. The investor group on climate change also has resources that provide information on the tools and frameworks emerging to help investors assess and manage physical climate risk at both the portfolio and the asset level to accelerate the management of resilience across the Australian economy.

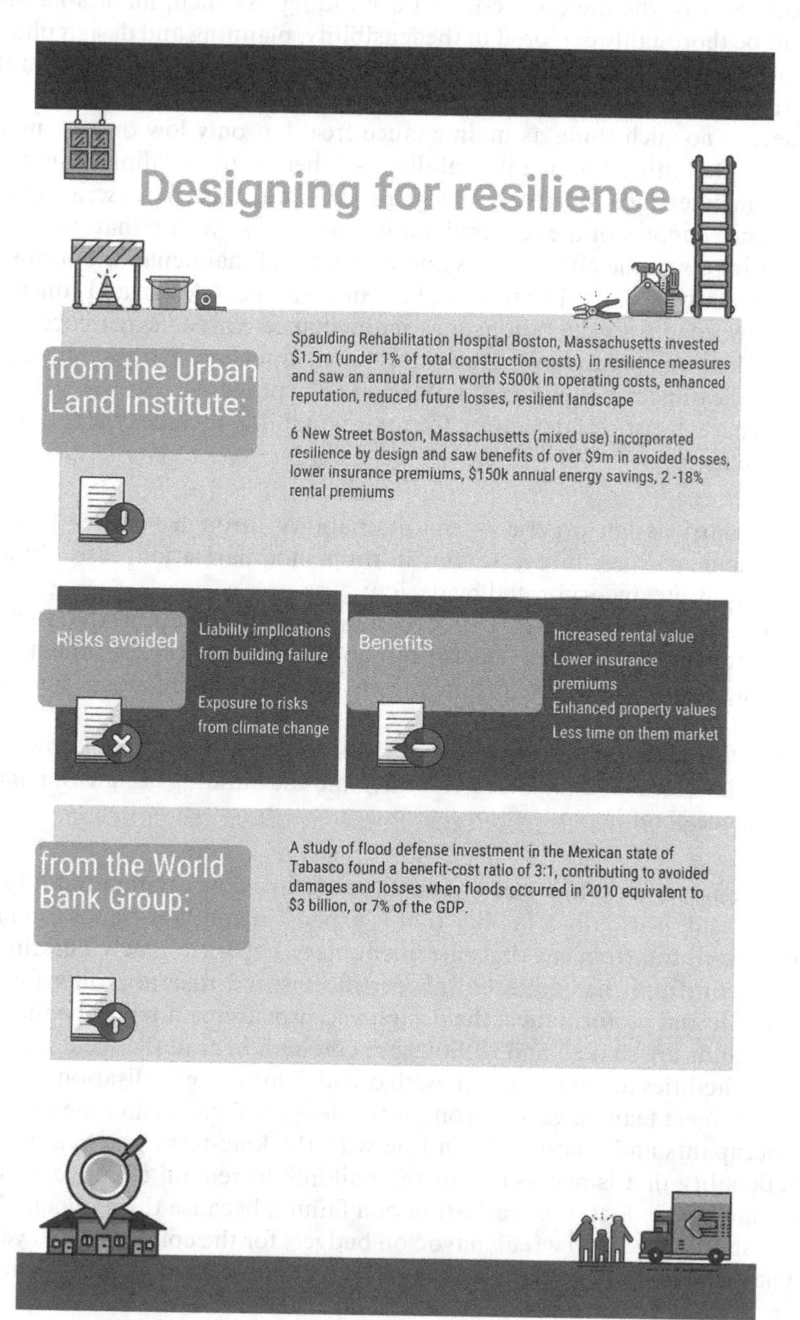

Maintenance

Operation, maintenance and repair costs amount to over three times the cost of initial construction. Numerous analyses indicate that this equates to 60%–80% of the life cycle costs of a building. As such, maintainability should be thoroughly explored in the feasibility, planning and design phases of a project rather than the standard practice of addressing it mainly in the construction and operations phase of a project.

There is no such thing as maintenance free, but only low or easy maintenance. Designing for maintainability is inherent to building design, by linking maintenance goals to the design process, it dictates the ease, accuracy, safety and economy of the required maintenance tasks within that system; as such, it improves the effectiveness and efficiency of maintenance optimising building life cycle costs. The timing of the integration of design and construction knowledge with operations and maintenance (O&M) experiences into project design is paramount. Designing for maintenance early significantly reduces the impact of equipment failure and downtime.

The characteristic of designing for maintainability includes the following among others:

- Standard design practice – maintainability through features such as equipment accessibility, standardisation, modularisation, ease of maintenance, etc. incorporated by design;
- Contract specifications – the development of specifications that include appropriately detailed operation and maintenance documentation along with any specific training needs, as well as maintenance management system requirements;
- Multidisciplinary involvement – incorporating input and knowledge sharing from maintenance personnel into the building design and maintenance planning.

When design choices are made without going through a maintainability filter, the result is usually a facility that is expensive and/or difficult to maintain. It is all too frequent that one encounters a sparkling new building; it looks beautiful; it has collected all certificates and rosettes going for design and initial performance; the design and procurement teams beam with pride; hands are shaken and ribbons are cut; and, behind the scenes, building and facilities managers groan with dread. This is the realisation that the overall project team have not brought the designers' vision and the needs of the occupants and community in line with the long-term practicality and functionality that is necessary for the building to remain effective over its life time. It is, in fact, the realisation of a failure; because this is when "hidden costs" come out to wreak havoc on budgets for the coming 50–60 years.

The following are some examples of where these "hidden costs" might arise:

- Insufficient access:
  - Ceiling lights that can only be accessed with scaffolding leading to extended down time and the need add the cost of hiring scaffolding to the budget for the life of the building. Alternatively, the lights will be completely abandoned when they fail;
  - Roof level ventilation and air conditioning systems with no lift access to transport parts such as filters or chemicals; ultimately this will mean that air quality or refrigeration will be compromised until something "bigger" fails like a fan that will force the maintenance to take place;
  - Fan coil units installed above false ceilings in cramped spaces, with no room for a ladder to be safely set up means that most likely they will not be properly maintained, or additional costs will be incurred to relocate them;
- Lack of layering so that equipment is installed easily during construction and becomes nearly inaccessible after final finishes are in place;
- New trees planted too close to new buildings leading to clogged rainwater drainage, risk to roof integrity when the tree grows and subsidence;
- Installing large glass facades without safe access for cleaning or replacement resulting in the need for expensive equipment such as cherry pickers or suspended aerial work platforms;
- Choosing components to install that are not easily or locally available and require long lead times for replacement leading to building downtime;
- Choosing components from a manufacturer known for not providing backwards compatibility meaning that any replacements will require more changes (such as fixings, valves pipe sizes, etc.) than otherwise necessary.

Considering the entire life of a building, it is 100s of times cheaper to invest time and some capital in good collaboration to allow the project team to account for buyer/owner requirements, energy efficiency, the skills of the designers, maintainability, needs of the developer/procurement team and capital costs. Meaningful dialogue is needed between engineers, architects, designers, buyers/occupants/developers, maintenance personnel and, once the job is awarded, contractors. The discussions must start at the first meeting and carry throughout the project. This is not often the case, capital budgets and timetables to be met persist in being the first priority with occupant needs and maintainability coming in at the bottom of the list. This is not helped by current market conditions or the certification schemes available which usually allow less than 10% of overall scores for maintainability. Furthermore, as soon as budgets are strained, value engineering is used and more recently it appears even while the design is in outline sketch form. Value engineering, at one time, used to refer to effective design with long-term costs taken into account; recently, however, it appears that value engineering refers to simply cutting costs, a serious misnomer if ever there was one.

When maintenance is taken into account consistently from the onset, life cycle costs can be reduced by up to 28%, building system downtime can be decreased by 30% and overall building operability increases by as much as 6%.

Thus, when assessing the BCR for building design decisions the following should be accounted for with respect to maintainability:

- Building shape layering and layout – will impact on access;
- Material choices across the board – impact on the type and frequency of maintenance needed and the availability of parts necessary throughout the life of the building;
- Servicing and control strategies – impact on the type and frequency of maintenance needed and the availability of parts necessary throughout the life of the building;
- Plant and equipment location – will impact on access;
- Maintenance planning – setting aside time and resources to establish maintainability goals, to develop of a policy statement setting these out and considering the appointment of a stakeholder champion to its development via specific project design reviews for maintainability;
- Knowledge share – allowing time for stakeholders impacted by maintenance practices to feedback on decision decisions at each stage but especially at any value engineering stages. This will include maintenance budget holders, facilities managers and maintenance personnel or contractors.

Social impact

In the UK, Europe and many other countries legislative requirements mean that social value has to be considered across a number of activities such as the award of service contracts by public bodies or through the reporting of social value contributions by private organisations. Building development presents an exceptional opportunity to add social value in its most fundamental and essential sense, beyond simple legislative requirements (such as levies or affordability quotas, etc.).

Common practices in assessing social value often do not take into account the wider value that a development could bring to the local community. For example, the most common added value quoted for a construction project is the opportunity it presents for new jobs. Standard analyses will quantify the number of jobs created and the value that the influx of new wages will have within the local economy. They will stop short, however, of assessing the value the job has to the individual and society as a whole. This means that viability is often underestimated and in many cases social value is undersold.

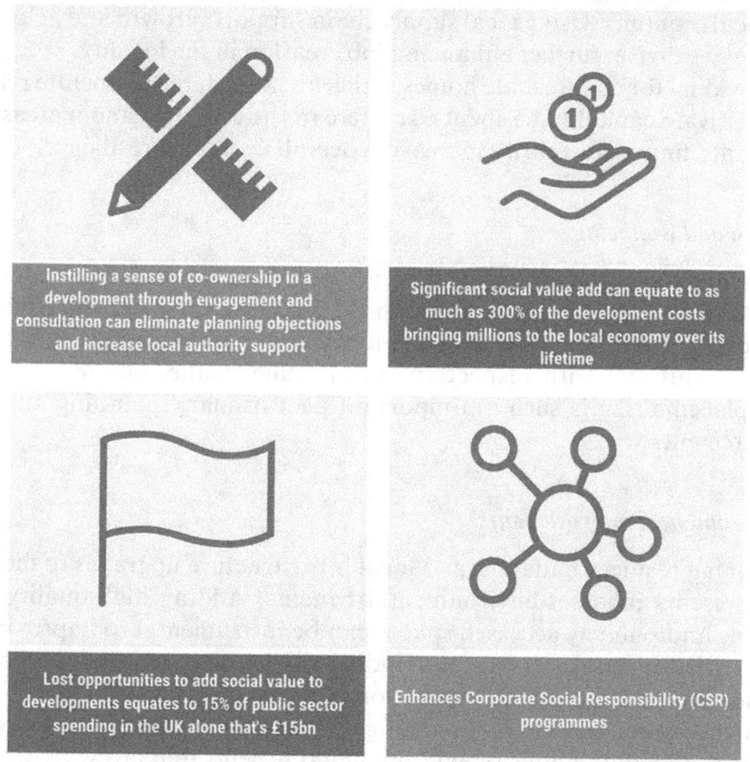

Primary to realising social value is recognising the importance of local context because the relative importance of different outcomes are dependent on the needs of the local area. Some potential outcomes that can be used as a starting point for pre-development BCR assessment and fall under the banner of social value are discussed in the following sections.

Job creation

Jobs can be created through the design with design team requiring specialist input from professionals, through construction, through the contractors with respect to hiring policies or through the building type such as office and retail units which offer local opportunities to local people. This helps meet national and international targets with respect to providing work for hard to reach groups and/or long-term employment and/or upskilling school leavers.

Economic growth

The economy benefits inherently from building projects. The level of materials and labour required to complete a building project sees to that. However, this can be maximised to provide benefits both to the developer team and to the local economy. Using local supply chain supports growth and helps local businesses thrive, further enhancing job creation in the locality.

Providing for comfortable homes which are affordable to operate releases more private capital to be spent elsewhere in the economy and increases individuals' financial security increasing overall economic resilience.

Health and wellbeing

Designing for occupants' health and wellbeing adds social value and should be assessed as such. Outwith the building's internal environment, improvements in the quality of the local urban and natural environment tend to multiply with respect to social value in the long-term. This is why placemaking is such an important part of many planning authority requirements.

Environmental improvements

Designing resilient buildings and local infrastructure upgrades to facilitate this presents inherent environmental benefits. Adding high-quality landscaping and publicly accessed spaces can be instrumental to improving the local environmental and increase biodiversity in the area as well as reducing the heat island effect which impacts on multiple surrounding spaces.

Furthermore, the acts of managing construction waste and resource use provides globally significant environmental benefits that last.

Community strength and safety

Local community groups frequently oppose development proposals when there is a perception that the development is not going to be delivered in the interests of the community. Communicating all aspects of added social value often helps eliminate these objections and smooths the progress of planning for the development. In itself, this can be the best reason to actively assess social value.

Aspects such as the effective and locally appropriate inclusion of affordable homes in a mixed use development provide strong local ownership of the development. Good accessibility and sustainable transportation are often missing in many locations and developments that can improve this are considered favourably. Construction practices such as the adoption of the considerate constructors scheme help protect existing social fabric from disruption.

When effective, developments have the potential to unlock derelict sites and strengthen community life therefore, community engagement and consultation is vitally important.

Notes

1 J. Fergus Nicol & Susan Roaf (2017) Rethinking thermal comfort, Building Research & Information, 45:7, 711–716, doi:10.1080/09613218.2017.1301698

2 Michael Pawlyn (2016) – design inspired by the way functional challenges have been solved in biology.

5 Whole life value

The incorporated approach

The practice of whole life valuation is intended for project teams that want to go beyond basic compliance or simply meeting legal requirements. It is an effective tool for presenting a good business case within the opportunities and constraints of real life. Using whole life value analysis allows us to develop the most cost efficient buildings and facilities taking into account the aspect of a building's realistic life span which, in many cases, is longer than that of its occupants.

There are over 70 free or proprietary resources and tools available publicly to help calculate whole life value and some of them have been mentioned in previous sections. Their appropriateness depends on the project type its stage and the priorities set at the beginning by decision makers. All of them help assess a particular aspect of the development.

Ultimately, the best tools are the ones that facilitate a multi-attribute decision analysis by accounting for non-financial measures to help prioritise financial measures when evaluating project alternatives.

The uniqueness of projects is so profound that one size never fits all. The following sections serve to help with customisation of any chosen set of tools using multi-attribute analysis techniques for making building project related decisions at appropriate times in the procurement process based on the design parameters, principles and benefits discussed thus far.

The basis for the analysis

Building projects have extended commonalities, and the differentiator is the importance of each of these. The most commonly quoted variables are listed below in no particular order.

- Location;
- Function;
- Aesthetics;
- Life span;
- Environmental performance;
- Flexibility;
- Occupancy;
- Maintenance;

- Financial performance;
- Timescale;
- Reliability;
- Security;
- Thermal performance;
- Wellbeing;
- Transportation;
- Digital performance.

Creating a hierarchy

At any given stage of the project process individual decision makers will allocate different hierarchies for different variables.

To illustrate the point here are two examples of hierarchical allocation.

A local authority is considering the rationalisation of multiple disparate offices across a city into a new purpose built iconic office building. It is in the process of identifying a suitable site, and the following are variables under consideration:

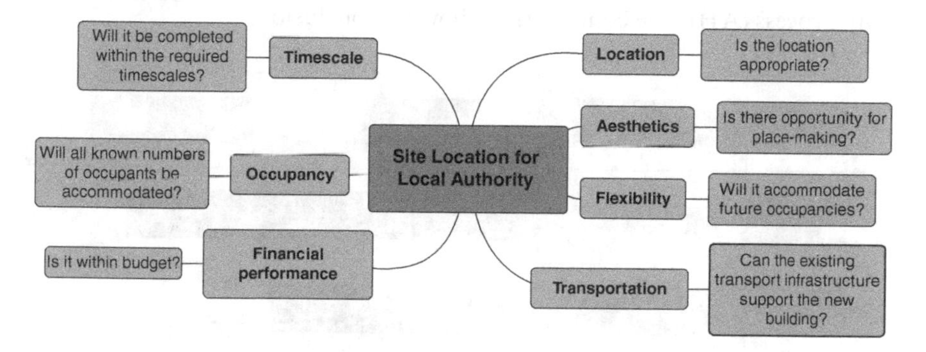

Assigning weightings and prioritising

The prioritisation of these variables will also depend on the individual decision makers. Let's assume the decisions are initially prioritised as follows:

Variable	*Description*	*Weighting based on importance*
1 Financial performance	Capital cost budgets	Highest – 100%
2 Occupancy	Whether the site fits a building big enough to house the known number of occupants	Highest – 100%
3 Location	Proximity to consolidated offices	High – 75%
4 Timescale	How quickly the building can get underway	High – 75%
5 Transportation	Public transport availability	Low – 50%
6 Aesthetics	Whether there is an opportunity for the resulting building to enhance the area by creating a sense of place	Low – 50%
7 Flexibility	Opportunity for future expansion	Lowest – 25%

Please note that for variables that are considered of no importance there is no need for them to be included at all.

Analysing options

There are three available sites (A, B and C) under consideration and they're characterised per variable as follows:

	1	*2*	*3*	*4*	*5*	*6*	*7*
A	Maybe	Maybe	Yes	Maybe	No	No	Yes
B	Yes	Yes	No	No	Yes	Yes	No
C	Maybe	No	Yes	Yes	Yes	Yes	Yes

Coming to a conclusion

By assigning further weightings to each of the answers as follows: Yes – 100%, Maybe – 50% and No – 0% and analysing using a standard analytical hierarchy process (AHP) we come to the following conclusion:

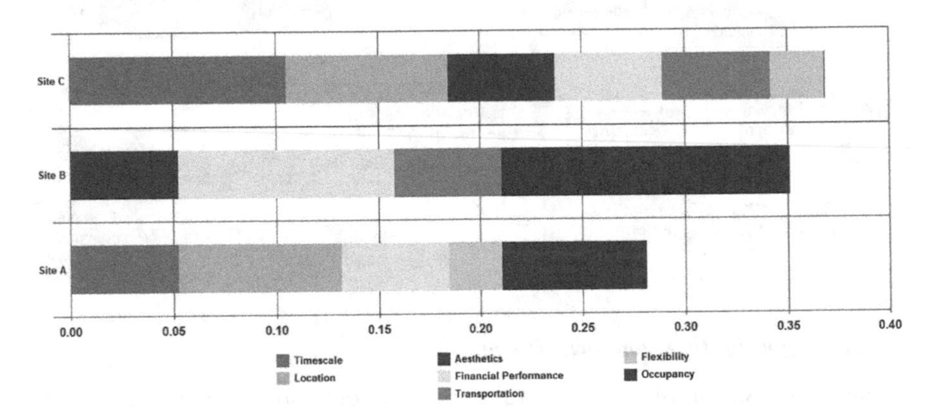

That site C is marginally more favourable than B and both are much more favourable than A. And, site C does have more going for it, it will be completed on time, there is sufficient transport infrastructure to get people to where they want to get it will be in close proximity to where all the previous offices were and it can be expanded if the need arises. However, it cannot accommodate all known employees. In this particular example taken from a real world example, the decision was made to implement flexible working patterns through shared desks and remote working for a percentage of the staff. This allowed a smaller building to be built. With respect to budget, it was realised that the "affordable" option was not as fully compliant with all parameters and so a new business case was made and accepted for additional funds to go with site C. It was felt that more funds could be released later down the construction process once agreement for the site purchase was reached. The example also indicates how changeable some decision processes are when all information is presented.

For the purposes of illustration one of the simplest decision making models on the market was used, DecisionPlus from Criterium. There are many other such tools available that provide equally effective results, and each project team will no doubt have favourites.

Basic background work

There is no right answer when it comes to prioritisation and deciding the hierarchy of factors to include in value calculations. The truth is that the project owner and their level of engagement will determine the ultimate success of the process. The rest of the project team and stakeholders will provide inputs but unless the weightings are allocated by the project owner the results will have no power to affect the outcome of the building project.

The project owner will need set or be assisted to set the overarching objectives for building performance, and the project team can use them as guidance to set priorities. Value charrettes are excellent ways to so this. Information and decisions on the aspects discussed in the following section will be required.

Project aim

The aim of the project will determine the overarching goal of any assessment. Some examples:

- To procure a good learning environment at as low a cost as possible;
- To deliver x domestic properties suitable for all stages of life within y budget by z year;
- To restore and repurpose a historic building to provide homes, retail and office space with stunning architecture whilst retaining the character and features of the existing building with the best energy efficiency possible.

Project values

Project values will help keep the project on track through the changeable process of design and construction and there needs to be at least five of them but more is preferable as they provide an insight into the real drivers for the project. Some examples:

- Value for money;
- High energy efficiency;
- Attracting higher quality tenants;
- Robustness – long building life;
- “Future proofed”;
- Productive learning/working environment;
- High occupancy ratios;
- High investment performance over the long term;
- Low operation costs, etc.

Interestingly, an Australian study on sustainable buildings indicated that 100% of survey respondents believed that sustainable buildings will outperform traditional buildings in the medium to long terms but not necessarily in the short term.

Project values need to be ranked in order of importance and associated weightings need to be allocated. As the project progresses, these values will be revisited and trivial matters will either be disregarded altogether or they may rise in importance. Conversely items that were critical to begin with may drop down the scale later on in the process depending on the inputs of other stakeholders. The weightings used in the examples here are what are termed verbal weightings associated with the following statements:

- Critical – 100%;
- Very important – 75%;
- Important – 50%;
- Unimportant – 25%;
- Trivial – 0%.

Key design parameters

The project owners need to decide on some overarching targets and design parameters that they term to be significant. This will inform the brief that the design team works to. Some examples include:

- Low environmental impact of materials;
- Low energy consumption in use – with numerical specification for what this should look like;
- High daylight levels – performance limits of how much daylight is expected and where;
- Adaptive thermal comfort – with clear performance specification of limits within which this will be considered acceptable;
- Increased biodiversity;
- High coverage for telecoms and IT infrastructure;
- Air quality – with specific numerical limits;
- Certification – with a target level of achievement;
- Accessibility;
- Ease of maintenance, etc.

Key financial parameters

Setting and making clear the total capital budget is crucial. If possible, it should also be split into Design and construction costs to help identify the value at stake in later stages. It should be noted that the cost of professional services (advisors, specialists, consultants, etc.) accounts for under 2% of the whole life cost is a project, yet the quality of these services has a direct impact on the other 98% and it really does not pay to be stingy on expertise.

Operational budgets will also be critical. Again, splitting them down to running cost and maintenance cost expectations will allow better decisions as the project progresses.

Preparation for future cost cutting will need to happen at this early stage as well. However, to a great extent the prioritisation of the various parameters discussed in the charrette will help aspects such as ring fencing of budgets and costs need to be confirmed.

Financial modelling agreements will be agreed at this stage too: how whole life costing will be carried out, which tools or standards will be used, when will reporting be required, by whom, to who, etc. Details of what will be included (e.g. acquisition, design, construction operation, maintenance and disposal) should be agreed at this point. The timing of the assessments needs also to be agreed.

Key stakeholders

Project stakeholders will need to be identified in full. Everyone that is involved in the building will have an input and will need to be included at appropriate times. The main stakeholders for most projects are:

Client side

- Investment decision maker – the person or committee in the organisation that decides whether investment should be made;
- Project owner – the person that requires the project with the status and authority to provide the necessary leadership;
- Project sponsor or client's representative– the person that provides the terms of reference, adequate staff, financial resources and support – they are the single focal point for the client's organisation in a project;
- Client advisor – usually a consultant appointed to assist non-technical project sponsors. They are not the project manager though; this is a more general role and this individual may also be in charge of facilitating decision making and whole value assessments;
- Specialist advisors – mostly external consultants that provide specialist advise directly to the project sponsor or owner; these include accessibility champions, sustainability champions, etc. they are tasked with monitoring and keeping track of the projects process with respect to their in individual specialisms and report back as needed;
- Project/client board – the stakeholders that the investment manager, the project owner and project sponsor will report to;
- Users – includes the expected functional users of the building as well as those that will be tasked with its operation and maintenance.

Project delivery side

- Project manager – the individual responsible for the day-to-day detailed management of the project providing an interface between the project's owner and sponsor and the supply side of the project team;

- Consultants – these are the professionals responsible for designing the building and include designers, cost consultants, specialist consultants such as health and safety, fire, building performance, acoustics commissioning, etc.;
- Contractors – the company/companies tasked with constructing the project they will include specialist subcontractors;
- Suppliers – the companies/individuals responsible for providing equipment, materials and all supplies for the project;

Others

- Authorities – these represent all interested parties with legislative powers over the project such as planners, building control, etc.;
- Local community – this includes all locally affected groups that will be impacted by the project and whose input will affect decision making.

Setting a communication strategy for these stakeholders is imperative. It may be that certain groups meet at certain stages or that others are required to report at specific intervals. The strategy will help dictate who and when makes decisions and how these are communicated.

Skill evaluation

People do not take their car to a greengrocer for its annual service and expect it to continue to be reliable and safe. In the same vein, buildings should be procured by people with appropriate skills.

Client side

On the project owners/client side it is important to understand the management abilities needed. Attributes include: decisiveness; communication; motivation; team building; facilitation; interviewing; negotiating; assertiveness; etc.

Technical competence is also required, knowledge of the following is critical and where it is not available in-house, serious consideration should be given to bringing it in for the project duration as an external advisory service:

- The construction industry;
- Value management;
- Whole life costing and multi-attribute evaluations;
- Procurement strategies;
- Risk management;
- Business case and investment appraisals;
- Project briefing;
- Contract strategies;
- Management planning;
- Estates strategies;

- Construction related legislation and regulations;
- Corporate social responsibility;
- Commissioning and building handover processes;
- Project financing;
- Claims.

Project delivery side

Technical competence is often implied when considering professionals and sometimes it is not challenged. The plethora of extended and expensive legal proceedings with respect to building projects indicate that an early stage skills assessment of the professionals involved in the project will at the very least help reduce the risk of this occurring. Some, and by no means all, experience and skill requirements are listed:

- Construction related legislation and regulations including further trends;
- Commissioning and building handover processes;
- Corporate social responsibility;
- Risk management;
- Value management;
- Whole life costing and multi attribute evaluations;
- Procurement strategies;
- Forms of contact;
- Project execution planning;
- Specification;
- Detailing;
- Design process;
- Design quality assessments;
- Energy management;
- Environmental issues;
- Space management;
- Site selection;
- Massing;
- Building form;
- Site planning;
- Security;
- Sustainability;
- Designing to operational targets;
- Embodied carbon;
- Climate change adaptation;
- Energy supplies;
- Energy minimisation;
- Building information modelling – to include operational energy, thermal comfort, indoor air quality, acoustics, occupant movements, daylight, parametric assessments, etc.;
- Collaborative working.

The importance of collaboration and engagement

Collaboration for the mutual benefit of all stakeholders should be the starting point on which relationships based equally between the client team and the project delivery team. Teamwork does not replace contractual agreements and appropriate management structures, but it is a pragmatic manner of working together to overcome the problems that will arise rather than continuing a blame culture that promotes wasteful activities and duplication of effort. It eliminates siloed working using collective knowledge and experience to find solutions. And, it significantly cuts down on nasty, costly and time-intensive consequences such as litigation, arbitration and dispute resolution.

Sample hierarchies

There are many aspects to consider for whole life valuation assessments. Total ownership costs including capital and cost in use are only the tip of the iceberg. Considerations touched upon in this book indicate that the attributes of a building that affect its perceived quality and effectiveness may feel like opening a can of worms. Obviously, not all will have the same importance or impact, and so some examples of what can be considered are included.

The detail with which they are assessed depends on the stage of the project. At earlier stages higher level attributes will need to be considered whereas later on more of the detail will be needed. This is why whole life value assessments cannot simply happen once on the design, rather they are needed at each important stage; they are live documents changing as the design evolves.

The main points of contact with the assessment process should be:

- Project inception;
- Feasibility appraisals;
- Planning and design;
- Value engineering;
- Construction and handover;
- Operation and maintenance;
- Decommissioning and renewal.

Decision making requires clear thinking on objectives and service requirements whenever it is undertaken and so whole value assessments and hierarchical structures for important project values need to be visited frequently.

Site selection

A sample hierarchy structure for the site selection stage is as presented here. The idea is to work with the primary project values (first level of criteria, five in total) and then expand slightly to include some outline design parameters (second level of criteria, 18 in total) to help put some context in the decision making.

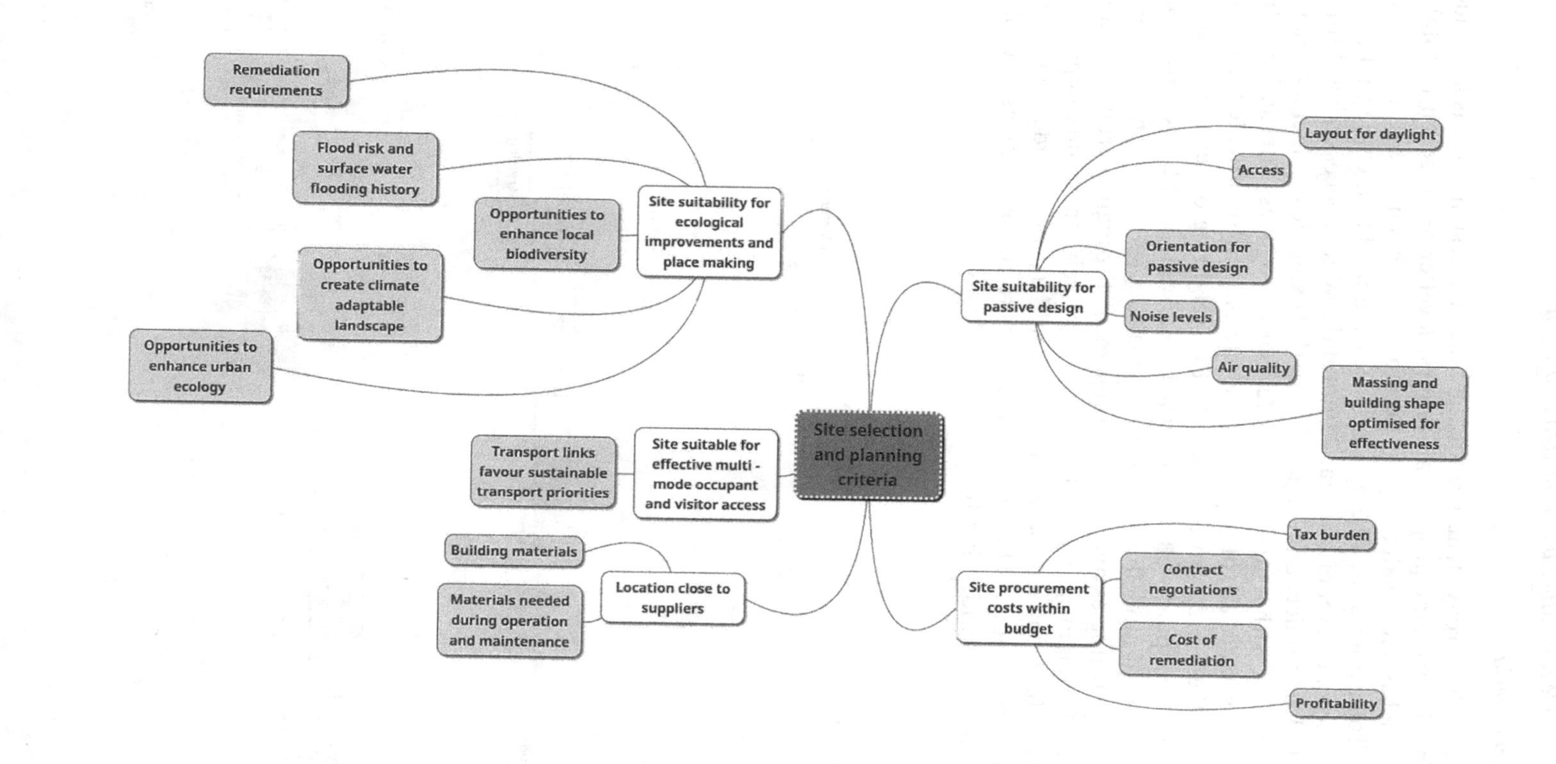

Site selection and planning criteria
Site suitability for ecological improvements and place making
Remediation requirements
Flood risk and surface water flooding history
Opportunities to enhance local biodiversity
Opportunities to create climate adaptable landscape
Opportunities to enhance urban ecology
Site suitable for effective multi - mode occupant and visitor access
Transport links favour sustainable transport priorities
Location close to suppliers
Building materials
Materials needed during operation and maintenance
Site suitability for passive design
Layout for daylight
Access
Orientation for passive design
Noise levels
Air quality
Massing and building shape optimised for effectiveness
Site procurement costs within budget
Tax burden
Contract negotiations
Cost of remediation
Profitability

Design criteria

A sample hierarchy structure for the building design stage is as included. Again, the primary project values (first-level criteria, seven in total) are used as a basis for evaluating the design and assessing the value of alternatives. To help with this these values are expanded to include slightly more detail (second level of criteria, 22 in total). Where necessary a third level is included to reflect any intricacies (in this example it is included in under maintenance with a further two criteria used to assess alternatives). In addition, some (not all) of the co-dependencies are also provided (connecting arrows); these are used to help with the weighting of certain aspects; for example, occupant wellbeing is shown to also depend by extension on energy (through modelling and servicing strategies, environmental through servicing strategies and space management through energy and building shape). If weighting is decided upon effectively, these interdependencies then can be deemed to be accounted for in the process and so omissions in design are avoided. This approach applies to the building as a whole and to components individually.

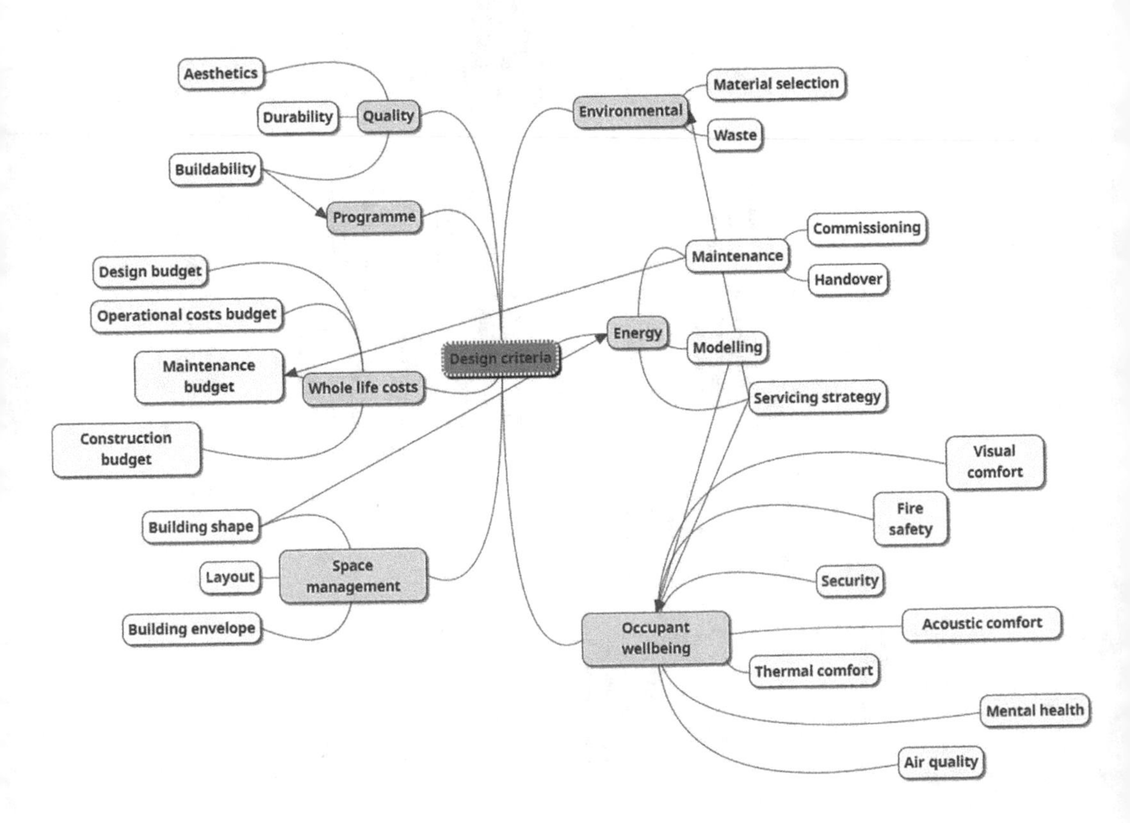

Material selection

A sample hierarchy structure for material selection is as depicted. Note how the design values are still present wherever they apply (five first-level criteria) and the 15 supporting second-level criteria provide the detail.

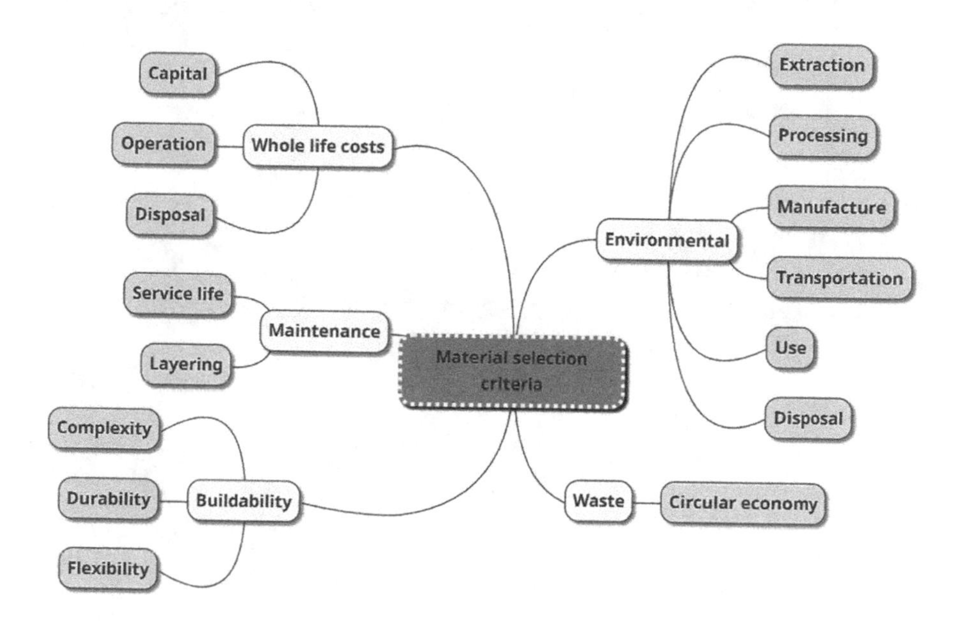

Repeatability

The process is fully repeatable and flexible enough to be adjusted as needed. It provides the transparency that is sometimes lost in the building design process and encourages collaboration at every stage it is applied to. The weighting and the assessment of individual criteria cannot be completed in silos and so communication and engagement is incorporated inherently when such practices are adopted.

Index

Taylor & Francis eBooks

www.taylorfrancis.com

A single destination for eBooks from Taylor & Francis with increased functionality and an improved user experience to meet the needs of our customers.

90,000+ eBooks of award-winning academic content in Humanities, Social Science, Science, Technology, Engineering, and Medical written by a global network of editors and authors.

TAYLOR & FRANCIS EBOOKS OFFERS:

A streamlined experience for our library customers

A single point of discovery for all of our eBook content

Improved search and discovery of content at both book and chapter level

REQUEST A FREE TRIAL

support@taylorfrancis.com